INTERNATIONAL ENCYCLOPEDIA of UNIFIED SCIENCE

Fundamentals of Concept Formation in Empirical Science

By

Carl G. Hempel

VOLUMES I AND II · FOUNDATIONS OF THE UNITY OF SCIENCE
VOLUME II · NUMBER 7

International Encyclopedia of Unified Science

Editor-in-Chief Otto Neurath†
Associate Editors Rudolf Carnap Charles Morris

Foundations of the Unity of Science

(Volumes I–II of the Encyclopedia)

Committee of Organization

Rudolf Carnap   Charles W. Morris
Philipp Frank†   Otto Neurath†
Joergen Joergensen   Louis Rougier

†Deceased.

The University of Chicago Press, Chicago 60637
The University of Chicago Press, Ltd., London

Copyright 1952 by The University of Chicago
All rights reserved. Published 1952
Eleventh Impression 1972
Printed in the United States of America
International Standard Book Number: 0-226-57597-7
Library of Congress Catalog Card Number: 52-13426

Contents:

	PAGE
1. Introduction	1
I. PRINCIPLES OF DEFINITION	2
2. On Nominal Definition	2
3. On "Real" Definition	6
4. Nominal Definition within Theoretical Systems	14
II. METHODS OF CONCEPT FORMATION IN SCIENCE	20
5. The Vocabulary of Science: Technical Terms and Observation Terms	20
6. Definition vs. Reduction to an Experiential Basis	23
7. Theoretical Constructs and Their Interpretation	29
8. Empirical and Systematic Import of Scientific Terms; Remarks on Operationism	39
III. SOME BASIC TYPES OF CONCEPT FORMATION IN SCIENCE	50
9. Classification	50
10. Classificatory vs. Comparative and Quantitative Concepts	54
11. Comparative Concepts and Nonmetrical Orders	58
12. Fundamental Measurement	62
13. Derived Measurement	69
14. Additivity and Extensiveness	75
NOTES	79
BIBLIOGRAPHY	88

Fundamentals of Concept Formation in Empirical Science

Carl G. Hempel

1. Introduction

Empirical science has two major objectives: to describe particular phenomena in the world of our experience and to establish general principles by means of which they can be explained and predicted. The explanatory and predictive principles of a scientific discipline are stated in its hypothetical generalizations and its theories; they characterize general patterns or regularities to which the individual phenomena conform and by virtue of which their occurrence can be systematically anticipated.

In the initial stages of scientific inquiry, descriptions as well as generalizations are stated in the vocabulary of everyday language. The growth of a scientific discipline, however, always brings with it the development of a system of specialized, more or less abstract, concepts and of a corresponding technical terminology. For what reasons and by what methods are these special concepts introduced and how do they function in scientific theory? These are the central questions which will be examined in this monograph.

It might seem plausible to assume that scientific concepts are always introduced by definition in terms of other concepts, which are already understood. As will be seen, this is by no means generally the case. Nevertheless, definition is an important method of concept formation, and we will therefore begin by surveying, in Chapter I, the fundamental principles of the general theory of definition. Chapter II will analyze the methods, both definitional and nondefinitional, by means of which scientific concepts are introduced. This analysis will lead to a closer examination of the function of concepts in scientific theories and will show that concept formation and theory forma-

Principles of Definition

tion in science are so closely interrelated as to constitute virtually two different aspects of the same procedure. Chapter III, finally, will be concerned with a study of qualitative and quantitative concepts and methods in empirical science.

We shall use, in this study, some of the concepts and techniques of modern logic and occasionally also a modicum of symbolic notation; these will, however, be explained, so that the main text of this monograph can be understood without any previous knowledge of symbolic logic. Some remarks of a somewhat more technical nature as well as points of detail and bibliographic references have been included in the notes at the end.[1]

I. Principles of Definition

2. On Nominal Definition

The word 'definition' has come to be used in several different senses. For a brief survey of the major meanings of the term, we choose as our point of departure the familiar distinction made in traditional logic between "nominal" and "real" definition. A real definition is conceived of as a statement of the "essential characteristics" of some entity, as when man is defined as a rational animal or a chair as a separate movable seat for one person. A nominal definition, on the other hand, is a convention which merely introduces an alternative—and usually abbreviatory—notation for a given linguistic expression, in the manner of the stipulation

(2.1) Let the word 'tiglon' be short for (i.e., synonymous with) the phrase 'offspring of a male tiger and a female lion'

In the present section we will discuss nominal definition; in the following one, real definition and its significance for scientific inquiry.

A *nominal definition* may be characterized as a stipulation to the effect that a specified expression, the *definiendum*, is to be synonymous with a certain other expression, the *definiens*, whose meaning is already determined. A nominal definition may therefore be put into the form

On Nominal Definition

(2.2)   Let the expression E_2 be synonymous with the expression E_1

This form is exemplified by the definition of the popular neologism 'tiglon' in (2.1) and by the following definitions of scientific terms:

(2.3) Let the term 'Americium' be synonymous with the phrase 'the element having 95 nuclear protons'

(2.4) Let the term 'antibiotic' be synonymous with (and thus short for) the expression 'bacteriostatic or bactericidal chemical agent produced by living organisms'

If a nominal definition is written in the form (2.2), it clearly speaks about certain linguistic expressions, which constitute its definiendum and its definiens; hence, it has to contain names for them. One simple and widely used method of forming a name for an expression is to put the expression between single quotation marks. This device is illustrated in the preceding examples and will frequently be used throughout this monograph.

There exists, however, an alternative way of formulating definitions, which dispenses with quotation marks, and which we will occasionally use. In its alternative form, the definition (2.3) would appear as follows:

(2.5)   Americium $=_{Df}$ the element with 95 nuclear protons

The notation '$=_{Df}$' may be read 'is, by definition, to equal in meaning', or briefly, 'equals by definition'; it may be viewed as stipulating the synonymy of the expressions flanking it. Here are two additional illustrations of this manner of stating nominal definitions:

(2.6)   the cephalic index of person x

$$=_{Df} 100 \frac{\text{maximum skull breadth of person } x}{\text{maximum skull length of person } x}$$

(2.7) x is dolichocephalic $=_{Df}$ x is a person with a cephalic index not exceeding 75

All these definitions are of the form

(2.8)   —————— $=_{Df}$ ·······

Principles of Definition

with the definiendum expression appearing to the left, and the definiens expression to the right of the symbol of definitional equality.

According to the account given so far, a nominal definition introduces, or defines, a new *expression*. But it is sometimes expeditious and indeed quite customary to describe the function of nominal definition in an alternative manner: We may say that a nominal definition singles out a certain *concept*, i.e., a nonlinguistic entity such as a property, a class, a relation, a function, or the like, and, for convenient reference, lays down a special name for it. Thus conceived, the definition (2.5) singles out a certain property, namely, that of being the chemical element whose atoms have 95 nuclear protons, and gives it a brief name. This second characterization is quite compatible with the first, and it elucidates the sense in which—as is often said—a nominal definition defines a *concept* (as distinguished from the expression naming it). Henceforth, we will permit ourselves to speak of definition, and later more generally of introduction, both in regard to expressions and in regard to concepts; the definition (2.6), for example, will be alternatively said to define the expression 'cephalic index of person x' or the concept of cephalic index of a person.

The expression defined by a nominal definition need not consist of just one single word or symbol, as it does in (2.5); it may instead be a compound phrase, as in (2.6) and (2.7). In particular, if the expression to be introduced is to be used only in certain specific linguistic contexts, then it is sufficient to provide synonyms for those contexts rather than for the new term in isolation. A definition which introduces a symbol s by providing synonyms for certain expressions containing s, but not for s itself, is called a *contextual definition*. Thus, e.g., when the term 'dolichocephalic' is to be used only in contexts of the form 'so-and-so is dolichocephalic', then it suffices to provide means for eliminating the term from those contexts; such means are provided by (2.7), which is a contextual definition.

The idea that the definiendum expression of an adequate nominal definition must consist only of the "new" term to be

On Nominal Definition

introduced is a misconception which is perhaps related to the doctrine of classical logic that every definition must be stated in terms of *genus proximum* and *differentia specifica*, as in the definition

(2.9)   minor $=_{Df}$ person less than 21 years of age

This definition characterizes, in effect, the class of minors as that subclass of the genus, persons, whose members have the specific characteristic of being less than 21 years old; in other words, the class of minors is defined as the logical product (the intersection) of the class of persons and the class of beings less than 21 years of age.

The doctrine that every definition must have this form is still widely accepted in elementary textbooks of logic, and it sometimes seriously hampers the adequate formulation of definitions—both nominal and "real"—in scientific writing and in dictionaries.[2] Actually, that doctrine is unjustifiable for several reasons. First, a definition by genus and differentia characterizes a class or a property as the logical product of two other classes or properties; hence this type of definition is inapplicable when the definiendum is not a class or a property but, say, a relation or a function. Consider, for example, the following contextual definition of the relation, harder than, for minerals:

(2.10) x is harder than $y =_{Df} x$ scratches y, but y does not scratch x

or consider the contextual definition of the average density of a body—which is an example of what, in logic, is called a function:

(2.11)   average density of $x =_{Df} \dfrac{\text{mass of } x \text{ in grams}}{\text{volume of } x \text{ in cc.}}$

In cases of this sort the traditional requirement is obviously inapplicable. And it is worth noting here that the majority of terms used in contemporary science are relation or function terms rather than class or property terms; in particular, all the terms representing metrical magnitudes are function terms and thus have a form which altogether precludes a definition by genus and differentia. Historically speaking, the genus-and-dif-

Principles of Definition

ferentia rule reflects the fact that traditional logic has been concerned almost exclusively with class or property concepts—a limitation which renders it incapable of providing a logical analysis of modern science.

But even for class or property concepts the traditional form of definition is not always required. Thus, e.g., a property might be defined as the logical sum of certain other properties rather than as a product. This is illustrated by the following definition, which is perfectly legitimate yet states neither genus nor differentia for the definiendum:

(2.12)   Scandinavian $=_{Df}$ Dane or Norwegian or Swede or Icelander

The genus-and-differentia form is therefore neither necessary nor sufficient for an adequate definition. Actually, the nominal definition of a term has to satisfy only one basic requirement: it must enable us to eliminate that term, from any context in which it can grammatically occur, in favor of other expressions, whose meaning is already understood. In principle, therefore, signs introduced by nominal definition can be dispensed with altogether: "To define a sign is to show how to avoid it."[3]

3. On "Real" Definition

A "real" definition, according to traditional logic, is not a stipulation determining the meaning of some expression but a statement of the "essential nature" or the "essential attributes" of some entity. The notion of essential nature, however, is so vague as to render this characterization useless for the purposes of rigorous inquiry. Yet it is often possible to reinterpret the quest for real definition in a manner which requires no reference to "essential natures" or "essential attributes," namely, as a search either for an empirical explanation of some phenomenon or for a meaning analysis. Thus, e.g., the familiar pronouncement that biology cannot as yet give us a definition of life clearly is not meant to deny the possibility of laying down some nominal definition for the term 'life'. Rather, it assumes that the term 'life' (or, alternatively, 'living organism') has a reasonably definite meaning, which we understand at least intuitively; and it asserts, in effect, that at present it is not possible to state, in a nontrivial manner, explicit and general criteria of life, i.e., conditions which are satisfied by just those phenomena which are instances of life according to the customary meaning of the term. A real definition of life would then consist in an equivalence sentence of the form

(3.1a) $\quad\quad x$ is a living organism if and only if
$\quad\quad\quad\quad x$ satisfies condition C

or, in abbreviatory symbolization:

(3.1b) $\quad\quad\quad\quad\quad Lx \equiv Cx$

Here, 'C' is short for an expression indicating a more or less complex set of conditions which together are necessary and sufficient for life. One set of conditions of this kind is suggested by Hutchinson[4] in the following passage:

> It is first essential to understand what is meant by a living organism. The necessary and sufficient condition for an object to be recognizable as a living organism, and so to be the subject of biological investigation, is that it be a discrete mass of matter, with a definite boundary, undergoing continual interchange of material with its surroundings without manifest alteration of properties over short periods of time, and, as ascertained either by direct observation or by analogy with other objects of the same class, originating by some process of division or fractionation from one or two pre-existing objects of the same kind. The criterion of continual interchange of material may be termed the *metabolic criterion*, that of origin from a pre-existing object of the same class, the *reproductive criterion*.

If we represent the characteristic of being a discrete mass with a definite boundary by 'D' and the metabolic and reproductive criteria by 'M' and 'R', respectively, then Hutchinson's characterization of life may be written thus:

(3.2) $\quad\quad\quad\quad Lx \equiv Dx \cdot Mx \cdot Rx$

i.e., a thing x is a living organism if and only if x has the characteristic of being a discrete mass, etc., and x satisfies the metabolic criterion, and x satisfies the reproductive criterion.

As the quoted passage shows, this equivalence is not offered as a convention concerning the use of the term 'living' but rather

Principles of Definition

as an assertion which claims to be true. How can an assertion of this kind be validated? Two possibilities present themselves:

The expression on the right-hand side of (3.2) might be claimed to be synonymous with the phrase 'x is a living organism'. In this case, the "real" definition (3.2) purports to characterize the meaning of the term 'living organism'; it constitutes what we shall call a *meaning analysis*, or an *analytic definition*, of that term (or, in an alternative locution, of the concept of living organism). Its validation thus requires solely a reflection upon the meanings of its constituent expressions and no empirical investigation of the characteristics of living organisms.

On the other hand, the "real" definition (3.2) might be intended to assert, not that the phrase 'x is a living organism' has the same meaning as the expression on the right, but rather that, as a matter of empirical fact, the three conditions D, M, and R are satisfied simultaneously by those and only those objects which are also living things. The sentence (3.2) would then have the character of an empirical law, and its validation would require reference to empirical evidence concerning the characteristics of living beings. In this case, (3.2) represents what we shall call an *empirical analysis* of the property of being a living organism.

It is not quite clear in which of these senses the quoted passage was actually intended; the first sentence suggests that a meaning analysis was aimed at.

Empirical analysis and meaning analysis differ from each other and from nominal definition. Empirical analysis is concerned not with linguistic expressions and their meanings but with empirical phenomena: it states characteristics which are, as a matter of empirical fact, both necessary and sufficient for the realization of the phenomenon under analysis. Usually, a sentence expressing an empirical analysis will have the character of a general law, as when air is characterized as a mixture, in specified proportions, of oxygen, nitrogen, and inert gases. Empirical analysis in terms of general laws is a special case

On "Real" Definition

of scientific *explanation*, which is aimed at the subsumption of empirical phenomena under general laws or theories.

Nominal definition and meaning analysis, on the other hand, deal with the meanings of linguistic expressions. But whereas a nominal definition introduces a "new" expression and gives it meaning by stipulation, an analytic definition is concerned with an expression which is already in use—let us call it the *analysandum expression* or, briefly, the *analysandum*—and makes its meaning explicit by providing a synonymous expression, the *analysans*, which, of course, has to be previously understood.

Dictionaries for a natural language are intended to provide analytic definitions for the words of that language; frequently, however, they supplement their meaning analyses by factual information about the subject matter at hand, as when, under the heading 'chlorine', a chemical characterization of the substance is supplemented by mentioning its use in various industrial processes.

According to the conception here outlined, an analytic definition is a statement which is true or false according as its analysans is, or is not, synonymous with its analysandum. Evidently, this conception of analytic definition presupposes a language whose expressions have precisely determined meanings—so that any two of its expressions can be said either to be, or not to be, synonymous. This condition is met, however, at best by certain artificial languages and surely is not generally satisfied by natural languages. Indeed, to determine the meaning of an expression in a given natural language as used by a specified linguistic community, one would have to ascertain the conditions under which the members of the community use—or, better, are disposed to use—the expression in question. Thus, e.g., to ascertain the meaning of the word 'hat' in contemporary English as spoken in the United States, we would have to determine to what kinds of objects—no matter whether they actually occur or not—the word 'hat' would be applied according to contemporary American usage. In this sense the conception of an analysis of "the" meaning of a given expression presupposes that the

Principles of Definition

conditions of its application are (1) well determined for every user of the language and are (2) the same for all users during the period of time under consideration. We shall refer to these two presuppositions as the conditions of *determinacy* and of (personal and interpersonal) *uniformity of usage*. Clearly, neither of them is fully satisfied by any natural language. For even if we disregard ambiguity, as exhibited by such words as 'field' and 'group', each of which has several distinct meanings, there remain the phenomena of vagueness (lack of determinacy) and of inconsistency of usage.[5] Thus, e.g., the term 'hat' is vague; i.e., various kinds of objects can be described or actually produced in regard to which one would be undecided whether to apply the term or not. In addition, the usage of the term exhibits certain inconsistencies both among different users and even for the same user of contemporary American English; i.e., instances can be described or actually produced of such a kind that different users, or even the same user at different times, will pass different judgments as to whether the term applies to those instances.

These considerations apply to the analysandum as well as to the analysans of an analytic definition in a natural language. Hence, the idea of a true analytic definition, i.e., one in which the meaning of the analysans is the same as that of the analysandum, rests on an untenable assumption. However, in many cases, there exists, for an expression in a natural language, a class of contexts in which its usage is practically uniform (for the word 'hat' this class would consist of all those contexts in which practically everybody would apply the term and of those in which practically none would); analytic definitions within a natural language might, therefore, be qualified as at least more or less adequate according to the extent to which uniform usage of the analysandum coincides with that of the analysans. When subsequently we speak of, or state, analytic definitions for expressions in a natural language, we will accordingly mean characterizations of approximately uniform patterns of usage.

Meaning analysis, or analytic definition, in the purely descriptive sense considered so far has to be distinguished from another procedure, which is likewise adumbrated in the vague traditional notion of real definition. This procedure is often called logical analysis or rational reconstruction, but we will refer to it, following Carnap's proposal, as *explication*.[6] Explication is concerned with expressions whose meaning in conversational language or even in scientific discourse is more or less vague (such as 'truth', 'probability', 'number', 'cause', 'law', 'explanation'—to mention some typical objects of explicatory study) and aims at giving those expressions a new and precisely determined meaning, so as to render them more suitable for clear and rigorous discourse on the subject matter at hand. The Frege-Russell theory of arithmetic and Tarski's semantical definition of truth are outstanding examples of explication.[7] The definitions proposed in these theories are not arrived at simply by an analysis of customary meanings. To be sure, the considerations leading to the precise definitions are guided initially by reference to customary scientific or conversational usage; but eventually the issues which call for clarification become so subtle that a study of prevailing usage can no longer shed any light upon them. Hence, the assignment of precise meanings to the terms under explication becomes a matter of judicious synthesis, of rational reconstruction, rather than of merely descriptive analysis: An explication sentence does not simply exhibit the commonly accepted meaning of the expression under study but rather proposes a specified new and precise meaning for it.

Explications, having the nature of proposals, cannot be qualified as being either true or false. Yet they are by no means a matter of arbitrary convention, for they have to satisfy two major requirements: First, the explicative reinterpretation of a term, or—as is often the case—of a set of related terms, must permit us to reformulate, in sentences of a syntactically precise form, at least a large part of what is customarily expressed by means of the terms under consideration. Second, it should be possible to develop, in terms of the reconstructed concepts, a comprehensive, rigorous, and sound theoretical system. Thus, e.g., the Frege-Russell reconstruction of arithmetic gives a clear

Principles of Definition

and uniform meaning to the arithmetical terms both in purely mathematical contexts, such as '7 + 5 = 12', and in their application to counting, as in the sentence 'The Sun has 9 major planets' (a purely axiomatic development of arithmetic would not accomplish this); and the proposed reconstruction provides a basis for the deductive development of pure arithmetic in such a way that all the familiar arithmetical principles can be proved.

Explication is not restricted to logical and mathematical concepts, however. Thus, e.g., the notions of purposiveness and of adaptive behavior, whose vagueness has fostered much obscure or inconclusive argumentation about the specific characteristics of biological phenomena, have become the objects of systematic explicatory efforts.[8] Again, the basic objective of the search for a "definition" of life is a precise and theoretically fruitful explication, or reconstruction, of the concept. Similarly, the controversy over whether a satisfactory definition of personality is attainable in purely psychological terms or requires reference to a cultural setting[9] centers around the question whether a sound explicatory or predictive theory of personality is possible without the use of sociocultural parameters; thus, the problem is one of explication.

An explication of a given set of terms, then, combines essential aspects of meaning analysis and of empirical analysis. Taking its departure from the customary meanings of the terms, explication aims at reducing the limitations, ambiguities, and inconsistencies of their ordinary usage by propounding a reinterpretation intended to enhance the clarity and precision of their meanings as well as their ability to function in hypotheses and theories with explanatory and predictive force. Thus understood, an explication cannot be qualified simply as true or false; but it may be adjudged more or less adequate according to the extent to which it attains its objectives.

In conclusion, let us note an important but frequently neglected requirement, which applies to analytic definitions and explications as well as to nominal definitions; we will call it the *requirement of syntactical determinacy:* A definition has to indicate the syntactical status, or, briefly, the syntax, of the expression it explicates or defines; i.e., it has to make clear the logical form of the contexts in which the term is to be used. Thus, e.g., the word 'husband' can occur in contexts of two different forms, namely, 'x is a husband of y' and 'x is a husband'. In the first type of context, which is illustrated by the sentence 'Prince Albert was the husband of Queen Victoria', the word 'husband' is used as a *relation term:* It has to be supplemented by two expressions referring to individuals if it is to form a sentence. In contexts of the second kind, such as 'John Smith is a husband', the word is used as a *property term*, requiring supplementation by only one individual name to form a sentence. Some standard English dictionaries, however, define the term 'husband' only by such phrases as 'man married to woman', which provide no explicit indication of its syntax but suggest the use of the word exclusively as a property term applying to married men; this disregards the relational use of the term, which is actually by far the more frequent. Similarly, the dictionary explication of 'twin' as 'being one of two children born at a birth' clearly suggests use of the word as a property term, i.e., in contexts of the form 'x is a twin', which is actually quite rare, and disregards its prevalent relational use in contexts of the form 'x and y are twins'. This shortcoming of many explications reflects the influence of classical logic with its insistence on construing all sentences as being of the subject-predicate type, which requires the interpretation of all predicates as property terms. Attempts to remedy this situation are likely to be impeded by the clumsiness of adequate formulations in English, which could, however, be considerably reduced by the use of variables. Thus, in a somewhat schematized form, an entry in the dictionary might read:

husband. (1) x is a h. of y: x is a male person, and x is married to y; (2) x is a h.: x is a male person who is married to some y.[10]

Nominal definitions have to satisfy the same requirement: Certainly, a term has not been defined if not even its syntax has been specified. In the definitions given in section 2, this condition is met by formulating the definiens in a way which unambiguously reflects its syntactical status; in some of them vari-

Principles of Definition

ables are used for greater clarity. Similarly the definition of a term such as 'force' in physics has to show that the term may occur in sentences of the form 'The force acting upon point P at time t equals vector f'. By way of contrast, consider now the concept of vital force or entelechy as adduced by neovitalists in an effort to explain certain biological phenomena which they consider as inaccessible in principle to any explanation by physicochemical theories. The term 'vital force' is used so loosely that not even its syntax is shown; no clear indication is given of whether it is to represent a property or a scalar or a vectorial magnitude, etc.; nor whether it is to be assigned to organisms, to biological processes, or to yet something else. The term is therefore unsuited for the formulation of even a moderately precise hypothesis or theory; consequently, it cannot possess the explanatory power ascribed to it.

A good illustration of the importance of syntactical determinacy is provided by the concept of probability. The definitions given in older textbooks, which speak of "the probability of an event" and thus present probabilities as numerical characteristics of individual events, overlook or conceal the fact that probabilities are relative to, and change with, some reference class (in the case of the statistical concept of probability) or some specific information (in the case of the logical concept of probability) and thus are numerical functions not of one but of two arguments. Disregard of this point is the source of various "paradoxes" of probability, in which "the same event" is shown to possess different probabilities, which actually result from a tacit shift in the reference system.

4. Nominal Definition within Theoretical Systems

Nominal definition plays its most important role in the formulation of scientific theories. In the present section we will consider the fundamental logical principles governing its use for this purpose.[11]

By the total vocabulary of a theory T let us understand the class of all the words or other signs which occur in the sentences of T. The total vocabulary of any scientific theory contains cer-

Nominal Definition within Theoretical Systems

tain terms which belong to the vocabulary of logic and mathematics, such as 'not', 'and', 'or', 'if . . . then ____', 'all', 'some', etc., or their symbolic equivalents; symbols for numbers as well as for operations on them and relations between them; and, finally, variables or equivalent verbal expressions. The balance of the terms in the total vocabulary of a theory T will be called the *extra-logical vocabulary*, or, briefly, the *vocabulary*, of T. Apart from a few exceptions which serve mainly illustrative purposes, we shall discuss here only the definition of extra-logical terms in scientific theories.

While many terms in the vocabulary of a theory may be defined by means of others, this is not possible for all of them, short of an infinite regress, in which the process of defining a term would never come to an end, or a definitional circle, in which certain terms would be defined, mediately or immediately, by means of themselves. Definitional circles are actually encountered in dictionaries, where one may find 'parent' defined by 'father or mother', then, 'father' in turn by 'male parent' and 'mother' by 'female parent'. This is unobjectionable for the type of analytic definition intended by dictionaries; in the context of nominal definition within scientific theories, however, such circularity is inadmissible because it defeats the purpose of nominal definition, namely, to introduce convenient notations which, at any time, can be eliminated in favor of the defining expressions. Infinite definitional regress evidently has to be barred for the same reason.

Thus the vocabulary of a theory falls into two classes: the *defined terms*, i.e., those which are introduced by definition in terms of other expressions of the vocabulary, and the so-called *primitive terms*, or *primitives*, by means of which all other terms of the theoretical vocabulary are ultimately defined. The primitives themselves, while not defined within the theory, may nevertheless have specific meanings assigned to them. Methods of effecting such interpretation of the primitives will be considered later.

By way of illustration let us define a set of words which might be used in a theory of family relationships. As primitives, we

Principles of Definition

choose the words 'male' and 'child'. The former will be used as a property term, i.e., in contexts of the form 'x is a male', or, briefly, 'Male x'; the latter will serve as a relation term, i.e., in contexts of the form 'x is a child of y', or, briefly, 'x Child y'. In formulating our definitions, we use, besides the dot symbol of conjunction, also the denial sign '$\sim$' (to be read 'it is not the case that'), and the notation for existential quantification—'$(Ez)(\ldots..)$' stands for 'there is at least one entity z such that ......'. As our universe of discourse, i.e., the totality of objects under consideration, we choose the class of human beings. Now we lay down the following definitions:

(4.1a) x Parent y $=_{Df}$ y Child x
(4.1b) x Father y $=_{Df}$ Male x · x Parent y
(4.1c) x Mother y $=_{Df}$ x Parent y · $\sim x$ Father y
(4.1d) x Grandparent y $=_{Df}$ $(Ez)(x$ Parent z · z Parent $y)$
(4.1e) x Grandmother y $=_{Df}$ $\sim$Male x · x Grandparent y

As the symbol '$=_{Df}$' indicates, these sentences are to be understood as nominal definitions, even though the meanings they assign to the definienda are those of ordinary usage if the primitives are taken in their customary meanings. Thus, e.g., the definition (4.1a) may be paraphrased as stipulating that 'x is a parent of y' is to be synonymous with 'y is a child of x', while (4.1d) lays down the convention that 'x is a grandparent of y' is to mean the same as 'x is a parent of someone, z, who is a parent of y'.

Let us note in passing that not one of these definitions has the genus-and-differentia form and, furthermore, that two of them, (4.1c) and (4.1e), are couched in what classical logic would call "negative terms." The traditional injunction against definition in negative terms[12] has no theoretical justification; indeed, it is highly questionable whether any precise meaning can be given to the very distinction of positive and negative concepts which it presupposes.

As the formulas (4.1) illustrate, each defined term in a theory is connected with the primitives by a "chain" of one or more definitions. This makes it possible to eliminate any occurrence

Nominal Definition within Theoretical Systems

of a defined term in favor of expressions in which all extra-logical symbols are primitives. Thus, 'parent' is eliminable directly in favor of 'child' by virtue of (4.1a); elimination of the word 'father' requires the use of two definitions; and the term 'grandmother' is linked to the definitional basis by a chain of three definitions, by virtue of which the phrase 'x Grandmother y' can always be replaced by the following expression, which contains only primitives: '$\sim$Male $x \cdot (Ez)(z$ Child $x \cdot y$ Child $z)$'. Generally, a *definition chain* for a term t, based on a given class of primitives, is a finite ordered set of definitions; in each of these any term occurring in the definiens either is one of the given primitives or has been defined in one of the preceding definitions of the chain; and the definiendum of the last definition is the term t.

Since any expression introduced by definition—i.e., by a single definition sentence or a chain of them—can be eliminated in favor of primitives, nominal definition, at least theoretically, can be entirely dispensed with: Everything that can be said with the help of defined terms can be said also by means of primitives alone. But even in our simple illustration the abbreviatory notations introduced by definition afford a noticeable convenience; and, in the complex theoretical systems of logic, mathematics, and empirical science, definitions are practically indispensable; for the formulation of those theories exclusively in terms of primitives would become so involved as to be unintelligible. Thus, not even the moderately advanced scientific disciplines could be understood—let alone actually have been developed—without extensive use of nominal definition.

The nominal definitions in a scientific theory are subject to one fundamental requirement, which we have mentioned repeatedly: They must permit the elimination of all defined terms in favor of primitives. More fully, this requirement may be stated thus:

Requirement of univocal eliminability of defined expressions:

For every sentence S containing defined expressions, there must exist an essentially unique expansion in primitive terms,

Principles of Definition

i.e., a sentence S' which satisfies the following conditions: (1) S' contains no defined term; (2) S' and S are deducible from one another with the help of the definition chains for the defined expressions occurring in S; (3) if S'' is another sentence which, in the sense of (2), is definitionally equivalent with S, then S' and S'' are logically deducible from each other and thus logically equivalent.

Thus, e.g., by virtue of the definition system (4.1), the phrase 'x Father y' has such alternative expansions as 'Male $x \cdot y$ Child x' and 'y Child $x \cdot$ Male x', but these are mutually deducible by virtue of the principles of formal logic.

The requirement of univocal eliminability has important consequences. First of all, it evidently precludes the possibility of giving two different definitions for the same term, an error which is usually avoided in practice but which could easily introduce contradictions if allowed to pass. In addition, a definition system which satisfies the requirement of univocal eliminability is noncircular, for any circularity would clearly preclude complete eliminability of defined terms.

It is sometimes held that nominal definitions, in contradistinction to "real" definitions, are arbitrary and may be chosen completely as we please. In reference to nominal definition in science, this characterization is apt to be more misleading than enlightening. For, in science, concepts are chosen with a view to functioning in fruitful theories, and this imposes definite limitations on the arbitrariness of definition, as will be pointed out in some detail in a later section. Furthermore, a nominal definition must not give rise to contradictions. As a consequence of this obvious requirement, the introduction of certain kinds of nominal definition into a given theoretical system is permissible only on condition that an appropriate nondefinitional sentence, which might be called its *justificatory sentence*, has been previously established. Thus, e.g., in Hilbert's axiomatization of Euclidean geometry, the line segment determined by two points, P_1 and P_2, is defined, in effect, as the class of points between P_1 and P_2 on the straight line through P_1 and P_2.[13] This definition evidently presupposes that through any

Nominal Definition within Theoretical Systems

two points there exists exactly one straight line; and it is permissible only because this presupposition can be proved in Hilbert's system and thus can function as justificatory sentence for the definition.

An illustration which was proposed and analyzed by Peano[14] shows well how disregard of the need for a justificatory theorem may engender contradictions. Consider the following definition of a "question-mark operation" for rational numbers:

$$\frac{x}{y} \mathbin{?} \frac{z}{u} =_{Df} \frac{x+z}{y+u}.$$

By virtue of this definition, we have:

$$\frac{1}{2} \mathbin{?} \frac{2}{3} = \frac{3}{5} \quad \text{and} \quad \frac{2}{4} \mathbin{?} \frac{2}{3} = \frac{4}{7};$$

but since $\frac{1}{2} = \frac{2}{4}$, it follows that $\frac{3}{5} = \frac{4}{7}$, which introduces a contradiction into arithmetic. Now, the given definition purports to introduce a question-mark operation as a unique function of its two arguments, i.e., in such a manner that to each couple of rational numbers—no matter in what particular form they are symbolized—it assigns exactly one rational number, which is to be regarded as the result of applying the question-mark operation to them. Clearly, this assumption is presupposed in deriving two incompatible results from the definition. But a definition purporting to introduce a unique function is acceptable only if accompanied by a justificatory theorem establishing this uniqueness—a requirement, which, in the case at hand, obviously cannot be met.

Consider now an illustration from the field of empirical science: The definition of the melting point of a given chemically homogeneous substance as the temperature at which the substance melts is permissible only if it has been previously established that all samples of that substance melt at the same temperature independently of other factors, such as pressure. Actually, the second of these conditions is not strictly satisfied, and in cases where the variation with pressure is marked a relativized concept of melting point at a specified pressure has to be used. Similar observations apply to the definitions of the density, specific heat, boiling point, specific resistance, and thermal conductivity of a substance, as well as to the definition of many other concepts in empirical science.

Nominal definitions in empirical science, then, are not entirely arbitrary, and in many cases they even require legitimation by a properly established justificatory sentence.

II. Methods of Concept Formation in Science

5. The Vocabulary of Science: Technical Terms and Observation Terms

Empirical science, we noted earlier, does not aim simply at a description of particular events: it looks for general principles which permit their explanation and prediction. And if a scientific discipline entirely lacks such principles, then it cannot establish any connections between different phenomena: it is unable to foresee future occurrences, and whatever knowledge it offers permits of no technological application, for all such application requires principles which predict what particular effects would occur if we brought about certain specified changes in a given system. It is, therefore, of paramount importance for science to develop a system of concepts which is suited for the formulation of general explanatory and predictive principles.

The vocabulary of everyday discourse, which science has to use at least initially, does permit the statement of generalizations, such as that any unsupported body will fall to the ground; that wood floats on water but that any metal sinks in it; that all crows are black; that men are more intellectual than women; etc. But such generalizations in everyday terms tend to have various shortcomings: (1) their constituent terms will often lack precision and uniformity of usage (as in the case of 'unsupported body', 'intellectual', etc.), and, as a consequence, the resulting statement will have no clear and precise meaning; (2) some of the generalizations are of very limited scope (as, for example, the statement dealing only with crows) and thus have small predictive and explanatory power (compare in this respect the

Technical Terms and Observation Terms

generalization about floating in water with the general statement of Archimedes' principle); (3) general principles couched in everyday terms usually have "exceptions," as is clearly illustrated by our examples.

In order to attain theories of great precision, wide scope, and high empirical confirmation, science has therefore evolved, in its different branches, comprehensive systems of special concepts, referred to by technical terms. Many of those concepts are highly abstract and bear little resemblance to the concrete concepts we use to describe the phenomena of our everyday experience. Actually, however, certain connections must obtain between these two classes of concepts; for science is ultimately intended to systematize the data of our experience, and this is possible only if scientific principles, even when couched in the most esoteric terms, have a bearing upon, and thus are conceptually connected with, statements reporting in "experiential terms" available in everyday language what has been established by immediate observation. Consequently, there will exist certain connections between the technical terms of empirical science and the experiential vocabulary; in fact, only by virtue of such connections can the technical terms of science have any empirical content. Much of the discussion in the present Chapter II will concern the nature of those connections. Before we can turn to this topic, however, we have to clarify somewhat more the notion of experiential term.

The experiential vocabulary is to be used in describing the kind of data which are usually said to be obtainable by direct experience and which serve to test scientific theories or hypotheses. Such experiential data might be conceived of as being sensations, perceptions, and similar phenomena of immediate experience; or else they might be construed as consisting in simple physical phenomena which are accessible to direct observation, such as the coincidence of the pointer of an instrument with a numbered mark on a dial; a change of color in a test substance or in the skin of a patient; the clicking of an amplifier connected with a Geiger counter; etc. The first of these two conceptions of experiential data calls for a phenomenological vocabulary, which might contain such expressions as 'blue-perception', 'looking brighter than' (applicable to areas of a visual field, not to physical objects), 'sour-taste-sensation', 'headachy feeling', etc. The second conception requires, for the description of experiential data, a set of terms signifying certain directly observable characteristics of physical objects, i.e., properties or relations whose presence or absence in a given case can be intersubjectively ascertained, under suitable circumstances, by direct observation. A vocabulary of this kind might include such terms as 'hard', 'liquid', 'blue', 'coincident with', 'contiguous with', etc., all of which are meant here to designate intersubjectively ascertainable attributes of physical objects. For brevity, we will refer to such attributes as *observables*, and to the terms naming them as *observation terms*.

A phenomenalistic conception will appeal to those who hold that the data of our immediate phenomenal experience must constitute the ultimate testing ground for all empirical knowledge; but it has at least two major disadvantages: first, while many epistemologists have favored this view, no one has ever developed in a precise manner a linguistic framework for the use of phenomenalistic terms;[15] and, second, as has been pointed out by Popper,[16] the use of observation reports couched in phenomenalistic language would seriously interfere with the intended objectivity of scientific knowledge: The latter requires that all statements of empirical science be capable of test by reference to evidence which is public, i.e., which can be secured by different observers and does not depend essentially on the observer. To this end, data which are to serve as scientific evidence should be described by means of terms whose use by scientific observers is marked by a high degree of determinacy and uniformity in the sense explained in section 3. These considerations strongly favor the second conception mentioned above, and we will therefore assume, henceforth, especially in the context of illustrations, that the vocabulary used in science for the description of experiential evidence consists of observation terms. Nevertheless, the basic general ideas of the following discussion can readily be transferred to the case of an experiential vocabulary of the phenomenalistic kind.

6. Definition vs. Reduction to an Experiential Basis

We now turn to a consideration of the connections between the technical terms of science and its observational vocabulary —connections which, as we noted, must exist if the technical terms are to have empirical content. Since the scientist has to introduce all his special terms on the basis of his observational vocabulary, the conjecture suggests itself that the former are defined in terms of the latter. Whether this is the case or not cannot be ascertained, however, by simply examining the writings and the pronouncements of scientists; for most presentations of science fail to state explicitly just what terms are taken to be defined and what others function as primitives. In general, only definitions of special importance will be stated, others will be tacitly taken for granted. Furthermore, the primitive terms of one presentation may be among the defined ones of another, and the formulations offered by different authors may involve various divergences and inconsistencies. The task of analyzing the logical relations among scientific terms is, therefore, one of rational reconstruction as characterized in section 3. Its ultimate objective is the construction of a language which is governed by well-determined rules, and in which all the statements of empirical science can be formulated. For the purposes of this monograph, it is not necessary to enter into the details of the complex problem—which is far from a complete solution— of how a rational reconstruction of the entire system of scientific concepts might be effected; it will suffice here to consider certain fundamental aspects of such a reconstruction.

The conjecture mentioned in the preceding paragraph may now be restated thus: Any term in the vocabulary of empirical science is definable by means of observation terms; i.e., it is possible to carry out a rational reconstruction of the language of science in such a way that all primitive terms are observation terms and all other terms are defined by means of them.

Methods of Concept Formation in Science

This view is characteristic of the earlier forms of positivism and empiricism, and we shall call it the *narrower thesis of empiricism*. According to it, any scientific statement, however abstract, could be transformed, by virtue of the definitions of its constituent technical terms, into an equivalent statement couched exclusively in observation terms: Science would really deal solely with observables. It might well be mentioned here that among contemporary psychologists this thesis has been intensively discussed in reference to the technical terms of psychology; much of the discussion has been concerned with the question whether the so-called intervening variables of learning theory are, or should be, completely definable in terms of directly observable characteristics of the stimulus and response situations.[17]

Despite its apparent plausibility, the narrower empiricist thesis does not stand up under closer scrutiny. There are at least two kinds of terms which raise difficulties: disposition terms, for which the correctness of the thesis is at least problematic, and quantitative terms, to which it surely does not apply. We will now discuss the status of disposition terms, leaving an examination of quantitative terms for the next section.

The property term 'magnetic' is an example of a disposition term: it designates, not a directly observable characteristic, but rather a disposition, on the part of some physical objects, to display specific reactions (such as attracting small iron objects) under certain specifiable circumstances (such as the presence of small iron objects in the vicinity). The vocabulary of empirical science abounds in disposition terms, such as 'elastic', 'conductor of heat', 'fissionable', 'catalyzer', 'phototropic', 'recessive trait', 'vasoconstrictor', 'introvert', 'somatotonic', 'matriarchate'; the following comments on the term 'magnetic' can be readily transferred to any one of them.

Since an object may be magnetic at one time and nonmagnetic at another, the word 'magnetic' will occur in contexts of the form '(object) x is magnetic at (time) t', and a contextual definition (cf. sec. 2) with this expression as definiendum has to be sought. The following formulation—which is deliberately

Definition vs. Reduction to an Experiential Basis

oversimplified in matters of physical detail—might suggest itself:

(6.1) x is magnetic at $t =_{Df}$ if, at t, a small iron object is close to x, then it moves toward x

But the conditional form of the definiens, while clearly reflecting the status of the definiendum as a disposition concept, gives rise to irksome problems.[18] In formal logic the phrase 'if . . . then ___' is usually construed in the sense of material implication, i.e., as being synonymous with 'either not . . . or also ___'; accordingly, the definiens of (6.1) would be satisfied by an object x not only if x was actually magnetic at time t but also if x was not magnetic but no small iron object happened to be near x at time t.

This shows that if sentences of the form illustrated by (6.1) are to serve as definitions for disposition terms, the 'if . . . then ___' clause in the definiens requires a different interpretation, whose import may be suggested by using the subjunctive mood:

(6.2) x is magnetic at $t =_{Df}$ if, at t, a small iron object should be close to x, then that object would move toward x.

Surely, the subjunctive conditional phrase cannot be interpreted in the sense of the material conditional; but before it can be accepted as providing an adequate formulation for the definition of disposition terms, the meaning of the phrase 'if . . . then ___' in subjunctive clauses would have to be made explicit. This is a problem of great interest and importance, since the formulation of so-called counterfactual conditionals and of general laws in science calls for the use of 'if . . . then ___' in the same sense; but despite considerable analytic efforts and significant partial results, no fully satisfactory explication seems available at present,[19] and the formulation (6.2) represents a program rather than a solution.

An alternative way of avoiding the shortcomings of (6.1) has been suggested, and developed in detail, by Carnap.[20] It consists in construing disposition terms as introduced, not by defini-

Methods of Concept Formation in Science

tion, but by a more general procedure, which he calls *reduction*, it amounts to partial, or conditional, definition and includes the standard procedure of explicit definition as a special case.

We will briefly explain this idea by reference to the simplest form of reduction, effected by means of so-called bilateral reduction sentences. A bilateral reduction sentence introducing a property term 'Q' has the form

(6.3) $$P_1x \supset (Qx \equiv P_2x)$$

Here, 'P_1x' and 'P_2x' symbolize certain characteristics which an object x may have; these may be more or less complex but must be stated in terms which are already understood.

In a somewhat loose paraphrase, which however suggests the scientific use of such sentences, (6.3) may be restated thus:

(6.31) If an object x has characteristic P_1 (e.g., x is subjected to specified test conditions or to some specified stimulus), then the attribute Q is to be assigned to x if and only if x shows the characteristic (i.e., the reaction, or the mode of response) P_2

Now the idea that was to be conveyed by (6.1) may be restated in the following reduction sentence:

(6.4) If a small iron object is close to x at t, then x is magnetic at t if and only if that object moves toward x at t

In reduction sentences, the phrase 'if . . . then ____' is always construed as synonymous with 'not . . . or ____', and 'if and only if' is understood analogously; yet the difficulty encountered by (6.1) does not arise for (6.4): If no small iron object is close to x at t, then the whole statement (6.4) is true of x, but we cannot infer that x is magnetic at t.

A reduction sentence offers no complete definition for the term it introduces, but only a partial, or conditional, determination of its meaning; it assigns meaning to the "new" term only for its application to objects which satisfy specific "test conditions." Thus, e.g., (6.4) determines the meaning of 'magnetic at t' only in reference to objects which meet the test condi-

Definition vs. Reduction to an Experiential Basis

tion of being close to some small iron body at t; it provides no interpretation for a sentence such as 'object x is now magnetic, but there is no iron whatever in its vicinity'. Hence, terms introduced by reduction sentences cannot generally be eliminated in favor of primitives. There is one exception to this rule: If the expression, 'P_1x', in (6.3) is analytic, i.e., is satisfied with logical necessity by any object x whatever (which is the case, for example, if 'P_1x' stands for 'x is green or not green'), then the bilateral reduction sentence is equivalent to the explicit definition '$Qx \equiv P_2x$'; hence, it fully specifies the meaning of 'Qx' and permits its elimination from any context. This shows that reduction is actually a generalization of definition. To put the matter in a different way, which will be useful later: A set of reduction sentences for a concept Q lays down a necessary condition for Q and a sufficient one; but, in general, the two are not identical. A definition of Q, on the other hand, specifies, in the definiens, a condition which is both necessary and sufficient for Q.

The indeterminacy in the meaning of a term introduced by a reduction sentence may be decreased by laying down additional reduction sentences for it which refer to different test conditions. Thus, e.g., if the concept of electric current had been introduced previously, (6.4) might be supplemented by the additional reduction sentence:

(6.5) If x moves through a closed wire loop at t, then x is magnetic at t if and only if an electric current flows in the loop at t

The sentences (6.4) and (6.5) together provide criteria of application for the word 'magnetic' in reference to any object that satisfies the test condition of at least one of them. But, since the two conditions are not exhaustive of all logical possibilities, the meaning of the word is still unspecified for many conceivable cases. On the other hand, the test conditions clearly are not logically exclusive: both may be satisfied by one and the same object; and for objects of this kind the two sentences imply a specific assertion, namely: Any physical object which is near some small iron body and moves through a closed wire loop will generate a current in the loop if and only if it attracts the iron body. But this statement surely is not just a stipulation concerning the use of a new term—in fact, it does not contain the new term, 'magnetic', at all; rather, it expresses an empirical law. Hence, while a single reduction sentence may be viewed simply as laying down a notational convention for the use of the term it introduces, this is no longer possible for a set of two or more reduction sentences concerning the same term, because such a set implies, as a rule, certain statements which have the character of empirical laws; such a set cannot, therefore, be used in science unless there is evidence to support the laws in question.

To summarize: An attempt to construe disposition terms as introduced by definition in terms of observables encounters the difficulties illustrated by reference to (6.1). These can be avoided by introducing disposition terms by sets of reduction sentences. But this method has two peculiar features: (1) In general, a set of reduction sentences for a given term does not have the sole function of a notational convention; rather, it also asserts, by implication, certain empirical statements. Sets of reduction sentences combine in a peculiar way the functions of concept formation and of theory formation. (2) In general, a set of reduction sentences determines the meaning of the introduced term only partially.

Now, as was noted in section 4, even an explicit nominal definition may imply a nondefinitional 'justificatory' statement which has to be established antecedently if the definition is to be acceptable; thus, the first characteristic of introduction by reduction sentences has its analogue in the case of definition. But this is not true of the second characteristic; and it might seem that the partial indeterminacy of meaning of terms introduced by reduction sentences is too high a price to pay for a method which avoids the shortcomings of definitions such as (6.1). It may be well, therefore, to suggest that this second characteristic of reduction sentences does justice to what appears to be an important characteristic of the more fruitful technical terms of science; let us call it their *openness of meaning*. The concepts of magnetization, of temperature, of gravitational field, for example, were introduced to serve as crystallization points for the formulation of explanatory and predictive principles. Since the latter are to bear upon phenomena accessible to direct observation, there must be "operational" criteria of application for their constitutive terms, i.e., criteria expressible in terms of observables. Reduction sentences make it possible to formulate such criteria. But precisely in the case of theoretically fruitful concepts, we want to permit, and indeed count on, the possibility that they may enter into further general principles, which will connect them with additional variables and will thus provide new criteria of application for them. We would deprive ourselves of these potentialities if we insisted on introducing the technical concepts of science by full definition in terms of observables.[21]

7. Theoretical Constructs and Their Interpretation

A second group of terms which fail to bear out the narrower thesis of empiricism are the metrical terms, which represent numerically measurable quantities such as length, mass, temperature, electric charge, etc. The term 'length', for example, is used in contexts of the form 'the length of the distance between points u and v is r cm.', or briefly

(7.1) $$\text{length } (u, v) = r$$

Similarly, the term 'mass' occurs in contexts of the form 'the mass of physical body x is s grams', or briefly

(7.2) $$\text{mass } (x) = s$$

In the hypotheses and theories of physics, these concepts are used in such a way that their values—r or s, respectively—may equal any nonnegative number. Thus, e.g., in Newton's general law of gravitation, which expresses the force of the gravitational attraction between two physical bodies as a function of their masses and their distance, all these magnitudes are allowed to take any positive real-number value. The concept of length, therefore—and similarly that of mass, and any other metrical concept whose range of values includes some interval of the real-number system—provides for the theoretical distinction of an infinity of different possible cases, each of them corresponding to one of the permissible real-number values. If, therefore, the concept of length were fully definable in terms of observables, then it would be possible to state, purely in terms of observables, the meaning of the phrase 'length $(u, v) = r$' for each of the permissible values of r. But this cannot be done, as we will now argue in two steps.

First, suppose that we try to define the characteristic of having a length of r cm. as tantamount to some specific combination (expressible by means of 'and', 'or', 'not', etc.) of observable attributes. (In effect, this restricts the definiens to a molecular sentence in which all predicates are observation terms.) This is surely not feasible for every theoretically permissible value of r. For in view of the limits of discrimination in direct observation, there will be altogether only a finite, though large, number of observable characteristics; hence, the number of different complexes that can be formed out of them will be finite as well, whereas the number of theoretically permissible r-values is infinite. Hence, the assignment of a numerical r-value of length (or of any other measurable quantity) to a given object cannot always be construed as definitionally equivalent to attributing to that object some specific complex of observable characteristics.

Let us try next, therefore, to construe the assignment of a specified r-value to a given object as equivalent to a statement about that object which can be expressed by means of observation terms and logical terms alone. The latter may now include not only 'and', 'or', 'not', etc., but also the expressions 'all', 'some', 'the class of all things satisfying such and such a condition', etc. But even if definition in terms of observables is construed in this broad sense, the total number of defining expressions that can be formed from the finite vocabulary available is only denumerably infinite, whereas the class of all theoretically permissible r-values has the power of the continuum. Hence, a full definition of metrical terms by means of observables is not possible.

It might be replied, in a pragmatist or extreme operationist vein, that a theoretical difference which makes no observable difference is no significant difference at all and that therefore no metrical concept in science should be allowed to take as its value just any real number within some specified interval. But compliance with this rule would make it impossible to use the concepts and principles of higher mathematics in the formulation and application of scientific theories. If, for example, we were to allow only a discrete set of values for length and for temporal duration, then the concepts of limit, derivative, and integral would be unavailable, and it would consequently be impossible to introduce the concepts of instantaneous speed and acceleration and to formulate the theory of motion. Similarly, all the formulations in terms of real and complex functions and in terms of differential equations, which are so characteristic of the theoretically most powerful branches of empirical science, would be barred. The retort that all those concepts and principles are "mere fictions to which nothing corresponds in experience" is, in effect, simply a restatement of the fact that theoretical constructs cannot be definitionally eliminated exclusively in favor of observation terms. But it is precisely these "fictitious" concepts rather than those fully definable by observables which enable science to interpret and organize the data of direct observation by means of a coherent and comprehensive system which permits explanation and prediction. Hence, rather than exclude those fruitful concepts on the ground that they are not experientially definable, we will have to inquire what nondefinitional methods might be suited for their introduction and experiential interpretation.

Do reduction sentences provide such a method? The conjecture has indeed been set forth in more recent empiricist writings that every term of empirical science can be introduced, on the basis of observation terms, by means of a suitable set of reduction sentences.[22] Let us call this assertion the *liberalized thesis of empiricism*.

Methods of Concept Formation in Science

But even for this thesis difficulties arise in the case of metrical terms. For, as was noted in section 6, a set of reduction sentences for a term t lays down a necessary and a (usually different) sufficient condition for the application of t. Hence, suitable reduction sentences for the phrase 'length $(u, v) = r$' would have to specify, for every theoretically permissible value of r, a necessary and a sufficient condition, couched in terms of observables, for an interval (u, v) having a length of exactly r cm.[23] But it is not even possible to formulate all the requisite sufficient conditions; for this would mean the establishment, for every possible value of r, of a purely observational criterion whose satisfaction by a given interval (u, v) would entail that the interval was exactly r cm. long. That a complete set of such criteria cannot exist is readily seen by an argument analogous to the one presented earlier in this section in reference to the limits of full definability in observational terms.

The metrical concepts in their theoretical use belong to the larger class of *theoretical constructs*, i.e., the often highly abstract terms used in the advanced stages of scientific theory formation, such as 'mass', 'mass point', 'rigid body', 'force', etc., in classical mechanics; 'absolute temperature', 'pressure', 'volume', 'Carnot process', etc., in classical thermodynamics; and 'electron', 'proton', 'ψ function', etc., in quantum mechanics. Terms of this kind are not introduced by definitions or reduction chains based on observables; in fact, they are not introduced by any piecemeal process of assigning meaning to them individually. Rather, the constructs used in a theory are introduced jointly, as it were, by setting up a theoretical system formulated in terms of them and by giving this system an experiential interpretation, which in turn confers empirical meaning on the theoretical constructs. Let us look at this procedure more closely.

Although in actual scientific practice the processes of framing a theoretical structure and of interpreting it are not always sharply separated, since the intended interpretation usually guides the constructions of the theoretician, it is possible and in-

Theoretical Constructs and Their Interpretation

deed desirable, for the purposes of logical clarification, to separate the two steps conceptually.

A theoretical system may then be conceived as an uninterpreted theory in axiomatic form, which is characterized by (1) a specified set of primitive terms; these are not defined within the theory, and all other extra-logical terms of the theory are obtained from them by nominal definition; (2) a set of postulates— we will alternatively call them primitive, or basic, hypotheses; other sentences of the theory are obtained from them by logical deduction.[24]

As an example of a well-axiomatized theory which is of fundamental importance for science, consider Euclidean geometry. Its development as "pure geometry," i.e., as an uninterpreted axiomatic system, is logically quite independent of its interpretation in physics and its use in navigation, surveying, etc. In Hilbert's axiomatization,[25] the primitives of the theory are the terms 'point', 'straight line', 'plane', 'incident on' (signifying a relation between a point and a line), 'between' (signifying a relation between points on a line), 'lies in' (signifying a relation between a point and a plane), and two further terms, for congruence among line segments and among angles, respectively. All other terms, such as 'parallel', 'angle', 'triangle', 'circle', are defined by means of the primitives: the term 'parallel', for example, can be introduced by the following contextual definition:

(7.3) x is parallel to $y =_{Df} x$ and y are straight lines; there exists a plane in which both x and y lie; but there exists no point which is incident on both x and y

The postulates include such sentences as these: For any two points there exists at least one, and at most one, straight line on which both are incident; between any two points incident on a straight line there exists another point which is incident on that line; etc. From the postulates, the other propositions of Euclidean geometry are obtained by logical deduction. Such proof establishes the propositions as theorems of pure mathematical geometry; it does not, however, certify their validity for use in physical theory and its applications, such as the determination of distances between physical bodies by means of triangulation, or the computation of the volume of a spherical object from the length of its diameter. For no specific meaning is assigned, in pure geometry, to the primitives of the theory[26] (and, consequently, none to the defined terms either); hence, pure geometry does not express any assertions about the spatial properties and relations of objects in the physical world.

A physical geometry, i.e., a theory which deals with the spatial aspects of physical phenomena, is obtained from a system of pure geometry by giving the primitives a specific interpretation in physical terms. Thus, e.g., to obtain the physical counterpart of pure Euclidean geometry, points may be interpreted as approximated by small physical objects, i.e., objects whose sizes are negligible compared to their mutual distances (they might be pinpoints, the intersections of cross-hairs, etc., or, for astronomical purposes, entire stars or even galactic systems); a straight line may be construed as the path of a light ray in a homogeneous medium; congruence of intervals as a physical relation characterizable in terms of coincidences of rigid rods; etc. This interpretation turns the postulates and theorems of pure geometry into statements of physics, and the question of their factual correctness now permits—and, indeed, requires—empirical tests. One of these is the measurement made by Gauss of the angle-sum in a triangle formed by light rays, to ascertain whether it equals two right angles, as asserted by physical geometry in its Euclidean form. If the evidence obtained by suitable methods is unfavorable, the Euclidean form of geometry may well be replaced by some non-Euclidean version which, in combination with the rest of physical theory, is in better accord with observational findings. In fact, just this has occurred in the general theory of relativity.[27]

In a similar manner, any other scientific theory may be conceived of as consisting of an uninterpreted, deductively developed system and of an interpretation which confers empirical import upon the terms and sentences of the latter.[28] The term to which the interpretation directly assigns an empirical content

Theoretical Constructs and Their Interpretation

either may be primitives of the theory, as in the geometrical example discussed before, or may be defined terms of the theoretical system. Thus, e.g., in a logical reconstruction of chemistry, the different elements might be defined by primitives referring to certain characteristics of their atomic structure; then, the terms 'hydrogen', 'helium', etc., thus defined might be given an empirical interpretation by reference to certain gross physical and chemical characteristics typical of the different elements. Such an interpretation of certain defined terms of a system confers mediately, as it were, some empirical content also upon the primitives of the system, which have received no direct empirical interpretation. This procedure appears well suited also for Woodger's axiomatization of biology,[29] in which certain defined concepts, such as division and fusion of cells, permit of a more direct empirical interpretation than some of the primitives of the system.

An adequate empirical interpretation turns a theoretical system into a testable theory: The hypotheses whose constituent terms have been interpreted become capable of test by reference to observable phenomena. Frequently the interpreted hypotheses will be derivative hypotheses of the theory; but their confirmation or disconfirmation by empirical data will then mediately strengthen or weaken also the primitive hypotheses from which they were derived. Thus, for example, the primitive hypotheses of the kinetic theory of heat concern the mechanical behavior of the micro-particles constituting a gas; hence, they are not capable of direct test. But they are indirectly testable because they entail derivative hypotheses which can be formulated in certain defined terms that have been interpreted by means of such "macroscopic observables" as the temperature and the pressure of a gas.[30]

The double function of such interpretation of defined terms—to indirectly confer empirical content upon the primitives of the theory and to render its basic hypotheses capable of test—is illustrated also by those hypotheses in physics or chemistry which refer to the value of some magnitude at a space-time point, such as the instantaneous speed and acceleration of a particle; or the density, pressure, and temperature of a substance at a certain point: none of these magnitudes is capable of direct observation, none of these hypotheses permits of direct test. The connection with the level of possible experimental or observational findings is established by defining, with the help of mathematical integration, certain derived concepts, such as those of average speed and acceleration in a certain time interval, or of average density in a certain spatial region, and by interpreting these in terms of more or less directly observable phenomena.

A scientific theory might therefore be likened to a complex spatial network: Its terms are represented by the knots, while the threads connecting the latter correspond, in part, to the definitions and, in part, to the fundamental and derivative hypotheses included in the theory. The whole system floats, as it were, above the plane of observation and is anchored to it by rules of interpretation. These might be viewed as strings which are not part of the network but link certain points of the latter with specific places in the plane of observation. By virtue of those interpretive connections, the network can function as a scientific theory: From certain observational data, we may ascend, via an interpretive string, to some point in the theoretical network, thence proceed, via definitions and hypotheses, to other points, from which another interpretive string permits a descent to the plane of observation.

In this manner an interpreted theory makes it possible to infer the occurrence of certain phenomena which can be described in observational terms and which may belong to the past or the future, on the basis of other such phenomena, whose occurrence has been previously ascertained. But the theoretical apparatus which provides these predictive and postdictive bridges from observational data to potential observational findings cannot, in general, be formulated in terms of observables alone. The entire history of scientific endeavor appears to show that in our world comprehensive, simple, and dependable principles for the explanation and prediction of observable phenomena cannot be obtained by merely summarizing and inductively generalizing observational findings. A hypothetico-deductive-observational procedure is called for and is indeed followed in the more advanced branches of empirical science: Guided by his knowledge of observational data, the scientist has to invent a set of concepts—theoretical constructs, which lack immediate experiential significance, a system of hypotheses couched in terms of them, and an interpretation for the resulting theoretical network; and all this in a manner which will establish explanatory and predictive connections between the data of direct observation.

Is it possible to specify a generally applicable form in which the interpretive statements for a scientific theory can be expressed? Let us note, to begin with, that those statements are not, in general, tantamount to full definitions in terms of observables. We will state the reasons by reference to the physical interpretation of geometrical terms. First, some of the expressions used in the interpretation, such as 'light ray in a homogeneous medium', are not observation terms but at best disposition terms which can be partially defined through observables by means of chains of reduction sentences.[31] Second, even if all the terms used in interpreting geometry were accepted as observation terms, the interpretive statements still would not express conditions which are both necessary and sufficient for the interpreted terms; hence, they would not have the import of definitions. If, e.g., it were necessary and sufficient for a physical point to be identical with, or at least at the same place as, a pinpoint or an intersection of cross-hairs or the like, then many propositions of geometry would be clearly false in their physical interpretation; among them, for example, the theorem that between any two points on a straight line there are infinitely many other points. Actually, no geometrical theory is rejected in physics for reasons of this type; rather, it is understood that comparatively small physical bodies constitute only approximations of points in the sense of physical geometry. The term 'point' as used in theoretical physics is a construct and does not denote any objects that are accessible to direct observation.

But an interpretation is not even generally tantamount to a set of reduction sentences: The interpretation of a theoretical term may well make use of expressions which are introducible by means of a set of reduction sentences based on observation predicates; but those expressions again will, as a rule, be used not to specify necessary and sufficient conditions for the theoretical term in question but only to provide a partial assignment of empirical content to it. Consider the case mentioned before of a reconstruction of chemistry in which the elements are theoretically defined in terms of their atomic structure and empirically interpreted by reference to their gross physical and chemical characteristics: Some of the latter, such as solubility in specified solvents, malleability, chemical affinity, etc., have the character of disposition concepts rather than of observables, and, furthermore, the interpretation is applicable only to sufficiently large amounts of the substance in question, so that surely only a partial interpretation of the theoretical terms is achieved.

As a consequence, an interpreted scientific theory cannot be equivalently translated into a system of sentences whose constituent extra-logical terms are all either observation predicates or obtainable from observation predicates by means of reduction sentences; and a fortiori no scientific theory is equivalent to a finite or infinite class of sentences describing potential experiences.

Considerations of the kind here surveyed have led some authors to the opinion that the rules of interpretation for a scientific theory cannot be stated in precise terms at all, that they will always have to remain somewhat vague.[32] Others have suggested that the interpretation of theoretical terms may have to be put into the form of probability statements.[33] Thus, e.g., the key terms of psychoanalytic theory (which, to be sure, has never been stated quite explicitly and precisely and for which no axiomatization is available) receive an empirical interpretation by reference to free associations, reports on dreams, slips of tongue, pen, and memory, and other more or less directly ascertainable aspects of overt behavior. But a cautious reconstruction will have to treat observable clues of these kinds not as strictly necessary or as strictly sufficient conditions for certain hypothetical states or processes such as oedipal fixation or regression or transference, but rather as "indicators" which are tied to those hypothetical states by probability relations. Thus, the rules of interpretation for a psychoanalytic concept C might specify the probabilities of specified observable symptoms occurring in individuals who have the nonobservable characteristic C, and conversely the probabilities for the presence of C if such and such observable symptoms are present. Generally, a theoretical system might be said to have been given an empirical interpretation if rules of confirmation have been laid down in such a way that for every sentence S of the theory and for every evidence sentence E that can be formulated in terms of observation predicates (no matter whether it is factually true or false) the rules determine (1) to what extent E confirms S, or what probability E confers upon S; and conversely (2) what probability S gives to E.

This conception of interpretation is at present largely a program, however. Its realization requires the development of an adequate theory of the probability of hypotheses. Important steps in this direction have been taken, but the results are still the object of controversy.[34] A discussion of details would take us too far beyond the scope of our present survey.

8. Empirical and Systematic Import of Scientific Terms; Remarks on Operationism

A theoretical system without empirical interpretation is incapable of test and thus cannot constitute a theory of empirical phenomena; we shall say of its terms as well as of its concepts that they lack *empirical import*.

Neovitalism, for example, provides no interpretation for its key term, 'entelechy', or for terms definable by means of it, nor does it offer an indirect interpretation by formulating a system of general laws and definitions which connect the term 'entelechy' with other, interpreted, terms of the theory. Consequently, the concept of entelechy cannot serve the explanatory purpose for which it was intended; for a concept can have explanatory power only in the context of an interpreted theory. Thus, e.g., to say that the regularities of planetary motion can be explained by means of the concept of universal gravitation is an elliptic way of asserting that those regularities are explainable by means of the formal *theory* of gravitation, together with the customary interpretation of its terms.

Another illustration is provided by the use of the term 'purpose' in some teleological accounts of biological phenomena. Thus, when a certain form of mimicry is said to have the purpose of preserving a given species by protecting its members from their natural enemies, no direct or indirect interpretation of 'purpose' in terms of observables is provided; i.e., no criteria are laid down by means of which it is possible to test assertions about the purposes of biological phenomena and to decide, let us say, between the view just mentioned and the alternative opinion that mimicry has the purpose of bringing aesthetic variety into the animate world. The assertion that mimicry actually does protect the members of a given species to some extent is not suited as interpretation for the former view, for then the alternative idea would have to be construed as stating that mimicry actually does effect aesthetic variety; and thus both claims would turn out to be true—a result which surely does not agree with the spirit and the intentions of teleological arguments. Thus, when used in contexts such as these, the term 'purpose' and the statements in which it occurs lack empirical import; they cannot provide any theoretical understanding of the phenomena in question. This use of teleological terminology may be characterized as an illegitimate transfer from contexts of similar grammatical form, for which, however, an empirical interpretation is available, such as 'the safety valves of steam engines have the purpose of preventing explosions'. This sentence can be interpreted by reference to the intentions and beliefs of the designer, which can be ascertained, at least under favorable conditions, by various observational methods. No doubt it is the close linguistic analogy to cases of the latter type which creates the illusion of empirical import in teleological arguments of the kind considered before.

Insistence that no term of science can be significant unless it possesses an empirical interpretation is the basic tenet of the operationist school of thought, which has its origin in the methodological work of the physicist P. W. Bridgman,[35] and which has exerted a great influence also in psychology and the social sciences.

The basic idea of operationism is "the demand that the concepts or terms used in the description of experience be framed in terms of operations which can be unequivocally performed";[36] in other words, the requirement that there must exist, for the terms of empirical science, criteria of application couched in terms of observational or experimental procedure. The operational criteria for the application of the term 'length', for example, would consist in appropriate rules for the measurement of length. But the idea should not be restricted to quantitative terms. Thus, the operational criteria of application for the term 'diphtheria' might be formulated in terms of the various symptoms of diphtheria; these would include not only such symptoms as are ascertainable by the "operation" of directly observing the patient but also the results of bacteriological and other tests which call for such "operations" as the use of microscopes and the application of staining techniques.

The statement of such criteria of application for a given term is often referred to as operational definition. This terminology is misleading, however. For, first of all, the term 'operational' would seem to preclude any criteria which require simply direct observation without any manipulation; and, as our last example illustrates, this would mean an unwarranted limitation. And, second, insistence that every scientific concept be *defined* in "operational" terms is unduly restrictive: as we tried to show in the preceding two sections, it would disqualify, among others, the most powerful theoretical constructs.

An attempt is sometimes made to reconcile insistence on a specific operational interpretation for all scientific terms with the endorsement of highly abstract theoretical constructs by representing the introduction of the latter as involving, in addition to "physical" operations, also "mental," "verbal," and "paper-and-pencil" operations.[37] This idea is then extended even to purely mathematical concepts, which are said to be defined in terms of mental operations. Such a view, however, fails to distinguish the systematic from the psychological aspects of concept formation. Thus, e.g., the definition of mathematical terms requires no reference to mental operations, although mental operations are involved in the psychological processes of defining and using mathematical terms. Moreover, mental operations are involved as well in the use of terms such as 'entelechy' or 'absolute simultaneity of spatially separated events', which the operationist criterion is intended to rule out as devoid of empirical import. Hence, to countenance "mental operations" in the criteria of application for scientific terms is to open a back door to all the concepts which operationism was originally designed to bar from the vocabulary of science.

An alternative to the reliance on "mental operations" and to the conception of operational criteria as generally providing full *definitions* is suggested by reflection upon the rationale of the operationist approach, i.e., the consideration that scientific terms are to function in statements which are capable of objective test by reference to data furnished by direct observation. "Operational definition" was conceived as a means to insure their suitability for this purpose by providing criteria for the test of the statements in which the terms occur. Thus, e.g., if the expression 'mineral x is harder than mineral y' is "operationally defined" by 'a sharp point of a sample of mineral x scratches a smooth surface of a sample of mineral y', then criteria have been established for the intersubjective test of comparative judgments about hardness. But the testability of a theory does not require full definition of its constituent concepts in terms of observables: a partial empirical interpretation will suffice. This suggests a broadening of the concept of operational definition: In its widest sense, we may construe an operational "definition" of a set of terms, $t_1, t_2, \ldots, t_n$ in a scientific theory as an interpretation, by reference to observables, of $t_1, t_2, \ldots, t_n$ or of other expressions which are connected with $t_1, t_2, \ldots, t_n$ by definitions within the theory. Thus, e.g., those operational criteria of diphtheria which refer to more or less directly observable symptoms might be viewed as a partial interpretation of the term 'diphtheria' itself, while the criteria referring to microscopic findings afford a partial interpretation of such terms as 'Krebs-Loeffler bacillus', which are definitionally connected with the term 'diphtheria' within bacteriological theory. In the broadened interpretation here suggested, the basic principle of operationism is just another formulation of the empiricist requirement of testability[38] for the theories of empirical science. Indeed, there exists a close correspondence between the ways in which those two ideas have been gradually liberalized. The earlier insistence that each statement of empirical science should be fully verifiable or falsifiable by means of observational evidence has been modified in two respects: (1) by the recognition that a scientific hypothesis cannot, as a rule, be tested in isolation but only in combination with other statements—the test of a bacteriological hypothesis by means of staining techniques and microscopes, for example, presupposes various hypotheses of mechanics, optics, and chemistry—so that the criterion of testability has to be applied to comprehensive systems of hypotheses rather than to single statements; and (2) by replacing the overly rigid standard of complete verifiability or falsifiability by the more liberal requirement that a system of hypotheses must be capable of being more or less highly confirmed by observational evidence. Analogously, the idea that each scientific term should be capable of definition in terms of observables has to be broadened by (1) application to a whole system of terms, connected by laws and definitions within a theoretical network, and (2) replacement of the requirement of complete definition by that of partial interpretation, as discussed in the preceding section.

As was noted in section 5, science strives for objectivity in the sense that its statements are to be capable of public test with results that do not vary essentially with the tester. This requirement makes it imperative that the vocabulary used in the interpretation of scientific terms have a high determinacy and uniformity of usage. This consideration accounts for the tendency to replace such experiential criteria as the direct comparison of two adjoining plane surfaces in regard to brightness (as used in photometry) or the reliance on specific smells, tastes, and visual appearances, or on the soapy feel of a lye in chemical work, by the use of instruments; this reduces the requisite observational evidence to statements describing spatial and temporal coincidences or specific pointer readings—a reduction which affords a considerable increase in pragmatic determinacy.

Another manifestation of the quest for scientific objectivity is the concern of psychologists and social scientists with the so-called *reliability* of their "operational definitions." Consider, for example, the social status scale propounded by Chapin.[39] Its purpose is to rate American homes in regard to their social status on the basis of readily obtainable information concerning the presence of such items as hardwood floors, radios, etc., and their condition of repair, as well as the general cleanliness and orderliness of the living-room. Each possible finding on these different counts is assigned a certain positive or negative integer as its weight, and the social status rating of a home is then determined by a specified arithmetical procedure from the weights assigned to it by a trained investigator. Evidently, the objectivity of the ratings thus obtained depends upon the determinacy and uniformity with which different investigators apply the basic criteria determining the scale. The reliability of the scale is intended to express this consistency of usage in numerical terms. Chapin used two different measures of reliability: (1) the correlation between the scores obtained by one observer on two sucessive visits to the same set of homes (this reflects what we have called determinacy of usage), and (2) the correlation between the scores obtained by two different observers for the same set of homes (this reflects uniformity of usage).[40] While the reliability thus obtained was high, it was not perfect, especially when determined in the second manner. The interobserver differences here reflected seem to be largely attributable to the fact that, while most of the rating criteria are purely descriptive in character, there is at least one which requires the investigator to express in numerical terms his general impression of good taste in the appearance of the living-room: If that impression is bizarre, inharmonious, or offensive, for example, the score on this count is -4. It is to be expected that valuational expressions have a smaller uniformity of usage than descriptive ones; their use in specifying the meanings of scientific terms will tend to conflict, therefore, with the requirement of uniform intersubjective testability for the statements of science.

Another illustration of this point is provided by Ogburn's hypothesis of cultural lag.[41] In rough summary, this hypothesis asserts that certain aspects of the nonmaterial culture of a social group are dependent on, and adapt themselves to, the prevailing material culture; so that a change in the latter (for example, excessive depletion of forests) will bring about a corresponding change in the adaptive culture (for example, enactment of forest conservation laws). However, between the occurrence of corresponding changes there is a temporal lag of varying length. An estimate of the lag in any particular case requires a determination not only of the time when a change in the adaptive culture did take place in response to a material change but also of the time when it "should" have taken place to prevent serious "maladjustment." But as Ogburn himself has candidly pointed out, "one's notion of adaptation in some cases depends somewhat on one's attitude towards life, one's idea of progress, or one's religious beliefs."[42] This is exactly why hypotheses employing such valuational notions are not capable of uniform intersubjective test and thus lack that objectivity which is an indispensable prerequisite of scientific formulations.[43] The factual information which the hypothesis of cultural lag is intended to convey needs therefore restating in a nonvaluational terminology.[44] Analogous considerations seem to be applicable to the functional theory of culture,[45] which proposes to account for social institutions and cultural change by reference to specific social needs they satisfy. Some of the concepts used for this purpose appear to be valuational, others teleological in character; in order to ascertain the empirical import of functional analyses, it is therefore essential that those concepts be given an interpretation in nonnormative terms.

Notwithstanding the soundness of the insistence on operational interpretations for scientific terms, it must not be forgotten that good scientific constructs must also possess *theoretical, or systematic, import;* i.e., they must permit the establishment of explanatory and predictive principles in the form of general laws or theories. Loosely speaking, the systematic import of a set of theoretical terms is determined by the scope, the degree of factual confirmation, and the formal simplicity of the general principles in which they function.[46]

In the theoretically advanced stages of science these two aspects of concept formation are inseparably connected; for, as we saw, the interpretation of a system of constructs presupposes a network of theoretical statements in which those constructs occur. In the initial stages of research, however, which are characterized by a largely observational vocabulary and by a low level of generalization, it is possible to separate the questions of empirical and of systematic import; and to do so explicitly may be helpful for a clarification of some rather important methodological issues.

Concepts with empirical import can be readily defined in any number, but most of them will be of no use for systematic purposes. Thus, we might define the hage of a person as the product of his height in millimeters and his age in years. This definition is operationally adequate and the term 'hage' thus introduced would have relatively high precision and uniformity of usage; but it lacks theoretical import, for we have no general laws connecting the hage of a person with other characteristics.

The significance of systematic import is well illustrated by the various attempts which have been made to divide human beings into different types according to physical or psychological characteristics. There are many different ways in which such typological classifications can be achieved by means of "operationally significant" criteria, so that the corresponding type concepts undeniably have empirical import. Yet the resulting systems differ considerably in scientific fruitfulness; for, in typological as well as in any other classificatory systems, it is essential that the defining characteristics of each of the different classes should be empirically associated with a large number of other attributes, so that the members of any one of the different classes will exhibit clusters of empirically correlated features. In typological systems, for example, one of the prevalent objectives is the delimitation of different types by means of physical characteristics which are closely correlated both with other physical attributes and with certain groups of psychological traits. Classifications which meet this requirement are often called *natural classifications*, in contradistinction to so-called *artificial classifications*, which are characterized by the absence of regular connections between the defining characteristics and others which are logically independent of them. This distinction will be examined more fully in section 9.

Recent typological theories, especially that of Sheldon,[47] allow for gradations in the various physical and mental traits they deal with. This yields a theoretical arrangement of individuals in an ordering framework reminiscent of a co-ordinate system instead of a classification with sharp boundary lines. A fruitful ordering schema is one in which the traits whose gradations determine the order have a high correlation with bundles of other physical or psychological characteristics.[48]

In the contemporary methodological literature of psychology and the social sciences, the need for "operational definitions" is often emphasized to the neglect of the requirement of systematic import, and occasionally the impression is given that the most promising way of furthering the growth of sociology as a scientific discipline is to create a large supply of "operationally defined" terms of high determinacy and uniformity of usage,[49] leaving it to subsequent research to discover whether these terms lend themselves to the formulation of fruitful theoretical principles. But concept formation in science cannot be separated from theoretical considerations; indeed, it is precisely the discovery of concept systems with theoretical import which advances scientific understanding; and such discovery requires scientific inventiveness and cannot be replaced by the—certainly indispensable, but also definitely insufficient—operationist or empiricist requirement of empirical import alone.

In the literature of psychology and sociology the opinion is sometimes expressed or implied that when a term of conversational language—such as 'manual dexterity', 'introversion', 'submissiveness', 'intelligence', 'social status', etc.—is taken over into the scientific vocabulary and given a more precise interpretation by reference to specified tests or similar criteria, then it is essential that the latter be *"valid"* in the sense of providing a correct characterization of the feature to which the term refers in ordinary usage. But if that term has so far been used solely in prescientific discourse, then the only way of ascertaining whether a proposed set of precise criteria affords a "valid" gauge of the characteristic in question is to determine to what extent the objects satisfying the criteria coincide with those to whom the characteristic in question would be assigned in prescientific usage. And, indeed, various authors have adopted, as an index of the validity of a proposed testing procedure for a psychological or social characteristic, the correlation between the test scores of a group of subjects and the ratings which those subjects were given, for the characteristic in question, by acquaintances who judged them "intuitively," i.e., who applied the term under consideration according to its prescientific usage. But the "intuitive" use of those terms in conversational language lacks both determinacy and uniformity. It is therefore unwarranted to consider them as referring to clearly delimited and unambiguously specifiable characteristics and to seek "correct" or "valid" indicators or tests for the presence or absence of the latter.

It should be added, however, that in the contemporary psychological and sociological literature, at least two other concepts of validity are used, which are not subject to this kind of criticism.[50] These concepts are applied especially to numerical scales (as determined by specified criteria) for such attributes as the pitch of a tone or the intelligence of an individual. In effect, one or both of the following criteria are applied to such scales (not being equivalent, they determine two different concepts of validity): (1) A high correlation of the given scale with other scales designed to represent the same attribute, or with some other previously adopted criterion; (2) ability of the attribute which is "operationally defined" by the criteria for the given scale to enter into fruitful and simple theoretical connections with other characteristics in the same area of investigation.[51] Validity in the first sense is a relative matter: A scale, or the graded characteristic represented by it, can be assigned a specific validity only in reference to certain previously accepted criteria or test scales; and its validity may be high in reference to one such set, low in reference to another. This conception of validity does not, therefore, seem to be of great theoretical importance; its major significance lies perhaps in the fact that it adumbrates, by aiming at correlation with other scaled characteristics, the requirement of systematic import; and it is again this requirement which clearly constitutes the rationale for the second conception of validity.

To summarize and amplify: In its exploratory pretheoretical research, science will often have to avail itself of the vocabulary of conversational language with all its imperfections; but in the course of its growth it has to modify its conceptual apparatus so as to enhance the theoretical import of the resulting system and the precision and uniformity of its interpretation —without being hampered by the consideration of preserving and explicating the prescientific usage of conventional terms taken over into its vocabulary. Physics, for example, would not have attained its present theoretical strength if it had insisted on using such terms as 'force', 'energy', 'field', 'heat', etc., in a manner that was "valid" in the sense discussed above.[52] At present, in fact, the connection between the technical and the prescientific meaning of theoretical terms has become quite tenuous in many instances, but the gain achieved by this "alienation" has been an enormous increase in the scope, simplicity, and experiential confirmation of scientific theories. Indeed, it is largely a matter of historical accident, and partly one of convenience, that terms of conversational English are used in the formulation of abstract theories; specially created words or symbols—as they are in fact used to some extent in all theoretical disciplines—would serve the same theoretical purpose and might offer the practical advantage of forestalling various misconceptions to which the use of familiar conversational terms gives rise.[53]

An interest in theoretical import increasingly influences concept formation also in psychology and sociology; and it is to be expected that if a system of concepts is advanced within these disciplines which has a clear empirical interpretation and considerable theoretical strength, it will be adopted even if it should differ radically from the conventional concepts and cut across the groupings and patterns established by them. In fact, such changes in the conceptual framework are clearly reflected in some recent theoretical developments. One of these is the attempt to isolate, through factor analysis,[54] systems of primary mental characteristics. The primary factors evolved by this method do not, as a rule, correspond closely to the psychological characteristics commonly distinguished in prescientific discourse. But this is irrelevant for the strictly systematic objective of the procedure, namely, the representation of numerous psychological traits, each of which is characterized by specific tests, as specific compounds—more exactly, as linear functions—of a relatively small number of independent primary factors.[55] While in factor analysis, emphasis is placed largely on a concept system which affords economy of descriptive representation, there are other theoretical trends which are aimed mainly at the construction of systems with explanatory and predictive power; these trends are illustrated, among others, by recent behavioristic theories of learning[56] and by the psychoanalytic and related approaches to certain aspects of personality and of culture.

III. Some Basic Types of Concept Formation in Science
9. Classification

The preceding chapter dealt with general logical and methodological problems of concept formation in empirical science. In the present chapter we will examine three widely used specific types of scientific concept formation, namely, the procedures of classification, of nonmetrical ordering, and of measurement. In this section we briefly consider classification.

Classification

Generally speaking, a classification of the objects in a given domain D (such as numbers, plane figures, chemical compounds, galactic systems, bacteria, human societies, etc.) is effected by laying down a set of two or more criteria such that every element of D satisfies exactly one of those criteria. Each criterion determines a certain class, namely, the class of all objects in D which satisfy the criterion. And if indeed each object in D satisfies exactly one of the criteria, then the classes thus determined are mutually exclusive, and they are jointly exhaustive of D.

Thus, e.g., one customary anthropometric classification of human skulls is based on the following criteria, C_1 to C_5:

(9.1) C_1: The cephalic index, $c(x)$, of skull x is 75 or less; or, $c(x) \leq 75$; C_2: $75 < c(x) \leq 77.6$; C_3: $77.6 < c(x) \leq 80$; C_4: $80 < c(x) \leq 83$; C_5: $83 < c(x)$

The properties determined by these criteria are referred to, respectively, as (1) dolichocephaly, (2) subdolichocephaly, (3) mesaticephaly, (4) subbrachycephaly, and (5) brachycephaly.

In this case the requirements of exclusiveness and exhaustiveness are satisfied simply as a logical consequence of the determining criteria; for, by definition, the cephalix index of any skull is a positive number, and every positive number falls into exactly one of the five intervals referred to by the criteria (9.1). An analogous observation applies, in particular, to all those dichotomous classifications which are determined by some property concept and its denial, such as the division of integers into those which are and those which are not integral multiples of 2, of chemical compounds into organic and inorganic, of bacteria into Gram-positive and Gram-negative.

Of greater significance for empirical science, however, is the case where at least one of the conditions of exclusiveness and exhaustiveness is satisfied not simply as a logical consequence of the determining criteria but as a matter of empirical fact; for this indicates an empirical law and thus confers some measure of systematic import upon the classificatory concepts involved. Thus, e.g., the division of humans into male and female on the

Some Basic Types of Concept Formation in Science

basis of primary sex characteristics or of the animal kingdom into various species, and the classification of crystals as developed in crystallography, are not logically exhaustive; hence, to the extent that they are factually so, they possess some systematic import by virtue of laws to the effect that every object in the domain under consideration satisfies one of the determining criteria.

As was mentioned earlier, a distinction is frequently made between *natural* and *artificial classifications*. The former are sometimes said to be based on essential characteristics of the things under investigation and to group together objects which possess fundamental similarities, whereas the latter are viewed as groupings determined by superficial resemblances or external criteria. Thus, e.g., the taxonomic division of plants or of animals into orders, families, genera, and species by reference to phylogenetic criteria would be considered as determining a natural classification; their division into several weight classes according to the average weight of the fully grown specimens would usually be regarded as artificial.

But the notion of essential characteristic invoked here is too obscure to be acceptable as the determining criterion for this classification of classifications. Indeed, it seems impossible to speak significantly of the essential characteristics of an individual thing. Surely, no examination of a given object could establish any of its characteristics as essential; and the customary interpretation of an essential characteristic of a thing as one without which the thing would not be what it is would obviously qualify every characteristic of a thing as essential and would thus render the concept trivial.

The idea of essential characteristic can be given a clearer meaning when it is used in reference to kinds or classes of objects rather than individual objects, as when chemical transmutation is said to be an essential characteristic of radioactive elements or metabolic activity an essential element of living organisms. Statements of this sort have the form 'Attribute Q is an essential characteristic of things of kind P', which may be construed as asserting that the characteristic Q is invariably associated with the characteristic P, i.e., that the sentence 'Whatever has the characteristic P also has the characteristic Q' is true either on logical grounds or as a matter of empirical fact. But this concept of essentiality can be applied only relatively to some given characteristic P; it does not justify the notion that the objects in a given field of inquiry can be individually described as to their essential characteristics and can then be divided into groups forming a natural classification.

The rational core of the distinction between natural and artificial classifications is suggested by the consideration that in so-called natural classifications the determining characteristics are associated, universally or in a high percentage of all cases, with other characteristics, of which they are logically independent. Thus, the two groups of primary sex characteristics determining the division of humans into male and female are each associated —by general law or by statistical correlation—with a host of concomitant characteristics; this makes it psychologically quite understandable that the classification should have been viewed as one "really existing in nature"—as contrasted with an "artificial" division of humans according to the first letter in their given names, or even according to whether their weight does or does not exceed fifty pounds.

Again, the taxonomic categories of genus, species, etc., as used in biology, determine classes whose elements share various biological characteristics other than those defining the classes in question; frequently, the groupings they establish also reflect relations of phylogenetic descent. Thus, the concepts by means of which biology seeks to establish a "natural system" are definitely chosen with a view to attaining systematic, and not merely descriptive, import: "The devising of a classification is, to some extent, as practical a task as the identification of specimens, but at the same time it involves more speculation and theorizing,"[57] for a "natural system is . . . one which enables us to make the maximum number of prophecies and deductions."[58] Similarly, the determining characteristics in the classification of crystals according to the number, relative length, and angles of inclination of their axes are empirically associated with a variety of other physical and chemical characteristics. Analogous observations apply to the ingenious arrangement of the chemical elements according to the periodic system, whose governing principles enabled Mendeleev to predict the existence of several elements then missing in the system and to anticipate with great accuracy a number of their physical and chemical properties.

If a natural classification is thus construed as one whose defining characteristics have a high systematic import, then evidently the distinction between natural and artificial classifications becomes a matter of degree; furthermore, the extent to which a proposed classification is systematically fruitful and thus natural has to be ascertained by empirical investigation; and, finally, a particular classification may well prove "natural" for the purposes of biology; others may be fruitful for psychology or sociology,[59] etc., and each of them would presumably be of little use in some of these contexts.

10. Classificatory vs. Comparative and Quantitative Concepts

A classificatory concept represents, as we saw, a characteristic which any object in the domain under consideration must either have or lack; if its meaning is precise, it divides the domain into two classes, separated by a sharp boundary line. Science uses concepts of this kind largely, though not exclusively, for the description of observational findings and for the formulation of initial, and often crude, empirical generalizations. But with growing emphasis on a more subtle and theoretically fruitful conceptual apparatus, classificatory concepts tend to be replaced by other types, which make it possible to deal with characteristics capable of gradations. In contrast to the "either ... or" character of classificatory concepts, these alternative types allow for a "more or less": each of them provides for a gradual transition from cases where the characteristic it represents is nearly or entirely absent to others where it is very marked. There are two major types of such concepts which are used in science: comparative and quantitative concepts.[60] These will

Classificatory vs. Comparative and Quantitative Concepts

now be described briefly, in preparation for a closer analysis in the following sections.

The idea of more or less of a certain attribute may be expressed in quantitative terms, as when the classificatory distinction of hot, cold, etc., is replaced by the concept of temperature in degrees centigrade. Concepts such as length in centimeters, temporal duration in seconds, temperature in degrees centigrade, etc., will be called *quantitative* or *metrical* concepts or, briefly, *quantities:* they attribute to each item in their domain of applicability a certain real number,[61] the value of the quantity for that item. In addition to these so-called scalar quantities, whose values are single numbers, there exist other metrical concepts, each of whose values is a set of several numbers; among these are vectors, such as velocity, acceleration, force, etc. The basic problems of measurement concern the introduction of scalars, and we will therefore limit our discussion of quantitative concepts to these.

But conceptual distinctions as to more or less may also be made without any use of numerical values, as when the classificatory terms 'hard', 'soft', etc., are replaced, in mineralogy, by the expressions 'x is as hard as y' and 'x is less hard than y'—both of which are defined by means of the scratch test. These two concepts allow for a comparison of any two pieces of mineral in regard to their relative hardness, and they thus determine an ordering of all pieces of mineral according to increasing hardness; but they do not introduce or presuppose any numerical measure of hardness. We will say that they determine a comparative concept of hardness, and we will generally refer to a concept based on criteria of the kind here illustrated as a *comparative concept*.

It has often been held that the transition from classificatory to the more elastic comparative and quantitative concepts in science has been necessary because the objects and events in the world we live in simply do not exhibit the rigid boundary lines called for by classificatory schemata, but present instead continuous transitions from one variety to another through a series of intermediate forms. Thus, e.g., the distinctions between long and short, hot and cold, liquid and solid, living and dead, male and female, etc., all appear, upon closer consideration, to be of a more-or-less character, and thus not to determine neat classifications. But this way of stating the matter is, at least, misleading. In principle, every one of the distinctions just mentioned can be dealt with in terms of classificatory schemata, simply by stipulating certain precise boundary lines. Thus, e.g., we might, by definitional fiat, qualify the interval between two points as long or short according as, when put alongside some arbitrarily chosen standard, the given interval does or does not extend beyond the latter. This criterion determines a dichotomous division of intervals according to length. And by means of several different standard intervals, we may exhaustively divide all intervals into any not too large finite number of "length classes," each of them having clearly specified boundaries.

However—and this observation suggests a more adequate way of stating the crucial point—suitably chosen comparative or quantitative concepts will often prove so considerably superior for the purposes of scientific description and systematization that they will seem to reflect the very nature of the subject matter under study, whereas the use of classificatory categories will seem an artificial imposition. We now turn to a brief survey of the major advantages offered by those nonclassificatory concepts.[62] Some of the characteristics included in our list will be found exclusively or predominantly in concepts of the quantitative type.

a) By means of ordering or metrical concepts, it is often possible to differentiate among instances which are lumped together in a given classification; in this sense a system of quantitative terms provides a greater descriptive flexibility and subtlety. Thus, e.g., the essentially classificatory wind scale of Beaufort distinguished twelve wind strengths: calm, light air, light breeze, gentle breeze, moderate breeze, fresh breeze, etc., which are defined by such criteria as smoke rising vertically, ripples on the surface of the water, white caps on the waves, etc. A corresponding quantitative concept is that of wind speed in miles per hour, which plainly permits subtler differentiations and which, in addition, covers all possible instances, whereas the

Classificatory vs. Comparative and Quantitative Concepts

Beaufort classes are not necessarily mutually exclusive and exhaustive of all possibilities. Now, the descriptive subtlety of a classificatory schema may be enhanced by the construction of narrower subclasses; but this possibility does not change the basic fact that the number of distinctions must remain limited. Besides, such subdivision requires the introduction of new terms for the various cases to be distinguished—an inconvenience which is avoided by metrical concepts.

b) A characterization of several items by means of a quantitative concept shows their relative position in the order represented by that concept; thus, a wind of 30 miles per hour is stronger than one of 18 miles per hour. Qualitative characterizations, such as 'gentle breeze' and 'moderate breeze', indicate no such relationship. This advantage of quantitative concepts is closely related to that obtained by the use of numerals rather than proper names in naming streets and houses: numerals indicate spatial relationships which are not reflected by proper names.

c) Greater descriptive flexibility also makes for greater flexibility in the formulation of general laws. Thus, e.g., by means of classificatory terms, we might formulate laws such as this: "When iron is warm, it is gray; when it gets hot, it turns red; and when it gets very hot, it turns white," whereas with the help of ordering terms of the metrical type it is possible to formulate vastly more subtle and precise laws which express the energy of radiation in different wave lengths as a mathematical function of the temperature.

d) The introduction of metrical terms makes possible an extensive application of the concepts and theories of higher mathematics: General laws can be expressed in the form of functional relationships between different quantities. This is in fact the standard form of general laws in the theoretically most advanced branches of empirical science; and mathematical methods such as those of the calculus can be used in applying scientific theories formulated in terms of mathematical functions to concrete situations, for purposes of test, prediction, or explanation.

In the following sections we will examine methods for the introduction of comparative and then of quantitative concepts. The introduction of a scalar quantitative concept will also be referred to as the determination of a *scale of measurement* or of a *metrical scale*.

Following Campbell, we distinguish two major ways of introducing scalar quantities: *fundamental measurement*, which presupposes no prior metrical scales, and *derived measurement*, i.e., the determination of one metrical concept by means of others, such as the definition of density in terms of mass and volume, or of certain anthropological indices as specified functions of the distances between certain reference points in the human body.

The most important—and perhaps the only—type of fundamental measurement used in the physical sciences is illustrated by the fundamental measurement of mass, length, temporal duration, and a number of other quantities. It consists of two steps: first, the specification of a comparative concept, which determines a nonmetrical order; and, second, the metrization of that order by the introduction of numerical values. In the next two sections these two steps will be analyzed in some detail. As our basic illustration we choose the fundamental measurement of mass by means of a balance.[63] This procedure is applicable only to bodies of "medium" size; let D_1 be the class of those bodies. The first, nonmetrical, stage of fundamental measurement will be discussed in section 11; the metrization of the resulting orders will be examined in section 12.[64]

11. Comparative Concepts and Nonmetrical Orders

To establish a comparative concept of mass for the class D_1 of medium-sized bodies is to specify criteria which determine for any two objects in D_1 whether they have the same mass, and, if not, which of them has the smaller one. Similarly, a comparative concept of hardness for the class D_2 of mineral objects is determined by criteria which specify, for any two elements of D_2, whether they are of equal hardness, and, if not, which of them is less hard. By means of these criteria it must be possible to arrange the elements of the given domain in a serial kind of order, in which an object *precedes* another if it has a smaller

Comparative Concepts and Nonmetrical Orders

mass, hardness, etc., than another, whereas objects of equal mass, hardness, etc., *coincide*, i.e., share the same place.

To generalize now: A comparative concept with the domain of application D is introduced by specifying criteria of coincidence and precedence for the elements of D in regard to the characteristic to be represented by the concept; the relations C of coincidence and P of precedence must be so chosen as to arrange the elements of D in a quasi-serial order, i.e., in an array that is serial except that several elements may occupy the same place in it. This means that C and P have to meet certain conditions, which will now be stated, and which will provide a precise definition of the concept of quasi-serial order. In the following formulations, x, y, and z are meant to be any elements of D, i.e., any of the objects to which the comparative concept characterized by C and P is to be applicable.

(11.1a) C is transitive; i.e., whenever x stands in C to y and y stands in C to z, then x stands in C to z; or briefly: $(xCy \cdot yCz) \supset xCz$

(11.1b) C is symmetric; i.e., whenever x stands in C to y, then y stands in C to x; or briefly: $xCy \supset yCx$

(11.1c) C is reflexive; i.e., any object x stands in C to itself; or briefly: xCx

(11.1d) P is transitive

(11.1e) If x stands in C to y, then x does not stand in P to y; or briefly: $xCy \supset \sim xPy$. If this condition is satisfied, we will say that P is C-irreflexive

(11.1f) If x does not stand in C to y, then x stands in P to y or y stands in P to x; or briefly: $\sim xCy \supset (xPy \text{ v } yPx)$. If this condition is satisfied, we will say that P is C-connected

The need for these requirements is intuitively clear, and it will presently be illustrated by reference to the comparative concept of mass. Here we note only that the last two conditions jointly amount to the requirement that any two elements of D

Some Basic Types of Concept Formation in Science

must be comparable in regard to the attribute under consideration; i.e., they must either have it to the same extent, or one must have it to a lesser extent than the other. This clarifies further the strict meaning of the idea of comparative concept. We now define:

(11.2) Two relations, C and P, determine a *comparative concept*, or a *quasi-series*, for the elements of a class D if, within D, C is transitive, symmetric, and reflexive, and P is transitive, C-irreflexive, and C-connected[65]

We now return to our illustration. In formulating specific criteria for this case, we will use two abbreviatory phrases: of any two objects, x and y, in D_1, we will say that x *outweighs* y if, when the objects are placed into opposite pans of a balance in a vacuum, x sinks and y rises; and we will say that x *balances* y if under the conditions described the balance remains in equilibrium.

A quasi-serial order for the elements of D according to increasing mass may now be determined by the following stipulations, in which x, y, z, are always assumed to belong to D_1:

(11.3a) x coincides with y, or x has the same mass as y, if and only if x is either identical with y or balances y

(11.3b) x precedes y, or x has less mass than y, if and only if y outweighs x

The two relations thus defined satisfy the requirements for quasi-serial order: Coincidence is reflexive and symmetrical simply as a consequence of the definition (11.3a); similarly, precedence is C-irreflexive as a consequence of our definitions. The satisfaction of the remaining requirements is a matter of empirical fact. Thus, e.g., the two relations are transitive by virtue of two general laws: Whenever a body x outweighs y and y outweighs z, then x outweighs z, and analogously for the relation representing coincidence. These two general statements do not simply follow from our definitions; in fact, the second of them holds only with certain limitations imposed by such factors as the sensitivity of the balance employed.[66]

Comparative Concepts and Nonmetrical Orders

The criteria laid down in (11.3) thus determine a nonmetrical order. They enable us to compare any two objects in D_1 in regard to mass; they do not assign, to the individual elements of D_1, numerical values as measures of their masses.

In mineralogy a comparative concept of hardness is defined by reference to the scratch test: A mineral x is called harder than another mineral, y, if a sharp point of x scratches a smooth surface of y; x and y are said to be of equal hardness if neither scratches the other. These criteria, however, are not entirely adequate for the determination of a quasi-series, for the relation of scratching is not strictly transitive.[67]

The rank orders which play a certain role in the initial stages of ordering concept formation in psychology and sociology are likewise of a nonmetrical character. A number of objects—frequently persons—are ordered by means of some criterion, which may be simply the intuitive judgment of one or more observers; the sequence in which the given objects—let there be n of them —are arranged by the criterion is indicated by assigning to them the integers from 1 to n. Any other monotonically increasing sequence of numbers—no matter whether integers or not—would serve the same purpose, since the rank numbers thus assigned have merely the function of ordinal numbers, not of measures. The purely ordinal character of ranking is reflected also by the fact that the rank assigned to an object will depend not only on that object but also on the group within which it is ranked. Thus, the number assigned to a student in ranking his class according to height, say, will depend on the other members of the class; the number representing a measure of his height will not. The need to change rank numbers when a new element is added to the group is sometimes avoided by inserting fractional ranks between the integral ones. Thus, e.g., in the hardness scale of Mohs, certain standard minerals are assigned the integers from 1 to 10 in such a way that a larger number indicates greater hardness in the sense of the scratch test. Now, this test places lead between the standard minerals talc, of hardness 1, and gypsum, of hardness 2; and instead of assigning to lead the rank number 2 and increasing that of all the harder standard min-

Some Basic Types of Concept Formation in Science

erals by 1, lead is assigned the hardness 1.5. Generally, a substance which the scratch test places between two standard minerals is assigned the arithmetic mean of the neighboring integral values. This method is imperfect, since it may assign the same number to substances which differ in hardness according to the scratch test. Similarly, in Sheldon's typologies of physique and of temperament,[68] positions intermediate to those marked by the integral values 1, 2, . . . , 7 in any one of the three typological components are indicated by so-called half numbers, i.e., by adding $\frac{1}{2}$ to the smaller integer.

When the intuitive judgment of some "qualified" observers is used as a criterion in comparing different individuals in regard to some psychological characteristic, then those observers play a role analogous to that of the balance in the definition of the comparative concept of mass. The use of human instruments of comparison has various disadvantages, however, over the use of nonorganismic devices such as scales, yardsticks, thermometers, etc.: Interobserver agreement is often far from perfect, so that the "yardstick" provided by one observer cannot be duplicated; besides, even one and the same observer may show inconsistent responses. In addition, it appears that most of the concepts defined by reference to the responses of human instruments are of very limited theoretical import—they do not give rise to precise and comprehensive generalizations.

12. Fundamental Measurement

In this section we examine first fundamental measurement as used in physics; the concept of mass again serves as an illustration. Subsequently, alternative types of fundamental measurement are considered briefly.

Fundamental measurement in physics is accomplished by laying down a quasi-serial order and then metricizing it by a particular procedure, which will be explained shortly. What we mean by metricizing a quasi-series is stated in the following definition:

(12.1) Let C and P be two relations which determine a quasi-serial order for a class D. We will say that this order has been metricized if criteria have been specified which assign to each element x of D exactly one real number, $s(x)$, in such a manner that the following conditions are satisfied for all elements x, y of D:

(12.1a)   If xCy, then $s(x) = s(y)$

(12.1b)   If xPy, then $s(x) < s(y)$

Any function s which assigns to every element x of D exactly one real-number value, $s(x)$, will be said to constitute a *quantitative* or *metrical concept*, or briefly a *quantity* (with the domain of application D); and if s meets the conditions just specified, we will say that it *accords with* the given quasi-series. It follows readily, in view of (11.2), that if s satisfies the conditions (12.1a) and (12.1b), then it also satisfies their converses, i.e.,

(12.1c)   If $s(x) = s(y)$, then xCy

(12.1d)   If $s(x) < s(y)$, then xPy

In the fundamental measurement of mass, a metrical concept which accords with the comparative one specified in (11.3) may be introduced through the following stipulations, in which '$m(x)$' stands for 'the mass, in grams, of object x':

(12.2a) If x balances y, then let $m(x) = m(y)$

(12.2b) If a physical body z consists of two bodies, x and y, which have no part in common and together exhaust z, then the m-value of z is to be the sum of the m-values of x and y. Under the specified conditions let us call z a *join* of x and y and let us symbolize this by writing: $z = xjy$. Now our stipulation can be put thus: $m(xjy) = m(x) + m(y)$

(12.2c) A specific object k, the International Prototype Kilogram, is to serve as a standard and is to be assigned the m-value 1,000; i.e., $m(k) = 1,000$

Now, by (12.2a), any object that balances the standard k has the m-value 1,000; by (12.2b), any object that balances a total of n such "copies" of k has an m-value of $1,000n$; each one of

Some Basic Types of Concept Formation in Science

two objects that balance each other and jointly balance k has the m-value 500; etc. In this manner it is possible—within limits set by the sensitivity of balances—to give an "operational interpretation" to integral and fractional multiples of the m-value 1,000 assigned to the standard k. Because of the limited sensitivity of balances, the rational values thus interpreted suffice to assign a mass value, $m(x)$, to any object x in the domain D_1; thus, the objective of fundamental measurement for mass has been attained.[69] Let us note that the conditions (12.1a) and (12.1b) are both satisfied; the former is enforced through the stipulation (12.2a), whereas the validity of the latter reflects a physical law: If x precedes y in the sense of (11.3b), i.e., if y outweighs x, then, as a matter of physical fact, y can be balanced by a combination of x with a suitable additional object z. But then, by (12.2b), $m(y) = m(x) + m(z)$; hence,[70] $m(x) < m(y)$.

The fundamental measurement of certain other magnitudes, such as length, and electrical resistance, exhibits the same logical structure: A quasi-serial order for the domain D in question is defined by means of two appropriately chosen relations, P and C, and this order is then metricized. The crucial phase in determining a scalar function s that accords with the given quasi-series consists, in each of those cases of fundamental measurement, in selecting some specific mode of combining any two objects of D into a new object, which again belongs to D; and in stipulating that the s-value of the combination of x and y is always to be the sum of the s-values of the components.

For a somewhat closer study of this process, let us choose, as a nonspecific symbol for any such mode of combination, a small circle (rather than the often-used plus sign, which is apt to be confusing). Thus, 'xoy' will designate the object obtained by combining, in some specified way, the objects x and y. The particular mode of combination varies from case to case; in the fundamental measurement of mass, e.g., it is the operation j of forming an object, xjy, which has x and y as parts; this can be done, in particular, by joining x and y or by placing them together. In the fundamental measurement of length the basic

Fundamental Measurement

mode of combination consists in placing intervals marked on rigid bodies end to end along a straight line.

We now state the stipulations by which, in fundamental physical measurement, a quasi-series, determined by two relations, C and P, is metricized with the help of some mode of combination, o; again, x and y are to be any elements of D.

(12.3a) If xCy, then $s(x) = s(y)$

(12.3b) $s(x\mathrm{o}y) = s(x) + s(y)$

(12.3c) Some particular element of D, say b, is chosen as a standard and is assigned some positive rational number r as its s-value: $s(b) = r$

To give another illustration of the third point: In the fundamental measurement of length in centimeters, b is the interval determined by two marks on the International Prototype Meter; r is chosen as 100.

Now, just as the relations C and P must meet specific conditions if they are to determine a quasi-series, so the operation o, together with C and P, has to satisfy certain requirements if the assignment of s-values specified by (12.3) is to be unambiguous and in accordance with the standards laid down in (12.1). We will refer to these requirements as the *conditions of extensiveness*. In effect, they demand that the operation o together with the relations C and P obey a set of rules which are analogous to certain laws which are satisfied by the arithmetical operation $+$, together with the arithmetical relations $=$ and $<$ (less than) among positive numbers. To consider a specific example: By (12.3b), we have $s(x\mathrm{o}y) = s(x) + s(y)$, and $s(y\mathrm{o}x) = s(y) + s(x)$; hence, by virtue of the commutative law of addition, which states that $a + b = b + a$, we have $s(x\mathrm{o}y) = s(y\mathrm{o}x)$. If, therefore, (12.1c) is to hold, we have to require that for any x and y in D,

(12.4a) $x\mathrm{o}y \; C \; y\mathrm{o}x$

This, then, is one of the conditions of extensiveness. Because of its strict formal analogy to the commutative law of addition, we

Some Basic Types of Concept Formation in Science

shall say that (12.4a) requires the operation o to be commutative relatively to C.

In a similar fashion the following further conditions of extensiveness are seen to be necessary.[71] (They are understood to apply to any elements $x, y, z, \ldots$, of D. The horseshoe again symbolizes the conditional, '$\ldots \supset$ ——' meaning 'if $\ldots$ then ——'; the dot serves as the sign of conjunction and thus stands for 'and'; finally, the existential quantifier, '(Ez)' is to be read 'there exists a thing z [in D] such that'.)

(12.4b) $\quad x\mathrm{o}(y\mathrm{o}z)\ C\ (x\mathrm{o}y)\mathrm{o}z$ (o must be associative relatively to C)

(12.4c) $\quad (xCy \cdot uCv) \supset x\mathrm{o}u\ C\ y\mathrm{o}v$

(12.4d) $\quad (xCy \cdot uPv) \supset x\mathrm{o}u\ P\ y\mathrm{o}v$

(12.4e) $\quad (xPy \cdot uPv) \supset x\mathrm{o}u\ P\ y\mathrm{o}v$

(12.4f) $\quad xCy \supset x\ P\ y\mathrm{o}z$

(12.4g) $\quad xPy \supset (Ez)(y\ C\ x\mathrm{o}z)$

The arithmetic analogues of these conditions are obtained by replacing 'o', 'C', 'P', by '$+$', '$=$', '$<$', respectively. The resulting formulas are readily seen to be true for all positive numbers. Condition (12.4f) serves to exclude what might be called zero-elements (in the case of mass, for example, a zero-element would be a body whose combination with another body does not yield an object with greater mass than the latter); while such elements often are useful for theoretical purposes, their admission on the level of fundamental measurement would introduce complications.

The conditions stated so far are sufficient to make sure that the stipulations (12.3) assign no more than one s-value to any one element of D and that the s-values thus assigned satisfy the requirements (12.1a) and (12.1b). They are not sufficient, however, to guarantee that the stipulations (12.3) assign some s-value to every object in D. If, for example, D should contain no two elements that stand in C to each other, then clearly the procedure of fundamental measurement specified in (12.3) would

Fundamental Measurement

assign an s-value to no other element of D than that chosen as a standard. The kind of further requirement that o, C, and P have to meet is suggested by the circumstance that fundamental measurement gives rise to rational s-values exclusively. Let us state this point more explicitly:

(12.5) The assignment, by the rules (12.3), of an s-value to an object x in D is based on finding a certain number, n, of other objects, $y_1, y_2, \ldots, y_n$, in F such that

(12.5a) any two of the y's stand in C to each other

(12.5b) joining a suitable number, say, m, of the y's by the operation o yields an object that stands in C to the standard b

(12.5c) by joining a suitable number, say k, of the y's, it is possible to obtain an object that stands in C to x

If these conditions are satisfied for a given object x, then the stipulations (12.3) yield $s(x) = (k \cdot r)/m$, which is a rational number. (By virtue of requiring a count of the number of intermediary standards, represented by the y's, which "make up" the standard b and the object x, respectively, the type of fundamental measurement here considered may be said to reduce ultimately to counting.)

If, for a given object x in D, there exists a set of intermediary standard objects $y_1, y_2, \ldots, y_n$ which satisfy the conditions listed in (12.5), we will say that x is *commensurable* with b in D on the basis of C and o. If x is not thus commensurable with b, then the rules of fundamental measurement assign no s-value to it at all. To insure, therefore, the assignment, required in (12.1) of an s-value to every x in D, we would have to supplement the requirements (12.4) by the following addition:

(12.6) *Condition of commensurability:* Every x in D is commensurable with b on the basis of C and o

Now, the limitations of observational discrimination preclude the possibility that this condition should ever be found to be violated in the fundamental measurement of physical quanti-

Some Basic Types of Concept Formation in Science

ties. Nevertheless, theoretical considerations strongly militate against its acceptance; for it restricts the possible values of those quantities to rational numbers, whereas it is of great importance for physical theory that irrational values be permitted as well. We will return to this point in the next section.

Condition (12.6), therefore, is imcompatible with physical theory and has to be abandoned. This has the consequence, however, that the stipulations for fundamental measurement as here considered do not guarantee the assignment of an s-value to each element of D and hence do not provide a full definition of the quantity s. Rather, they have to be viewed as a partial interpretation,[72] by reference to observables, of the expression '$s(x)$', which itself has the status of a theoretical construct. In leading to this result, our discussion corroborates and amplifies the ideas presented in section 7; it also makes clear that even the fundamental measurement of physical quantities requires the satisfaction of various general conditions, namely, those for quasi-serial order and for extensiveness. And while, depending on the particular choice of D, C, P, o, and b, some of these conditions may be true just by definition, others will have the character of empirical laws. Hence, fundamental measurement is not just a matter of laying down certain operational rules: it must go hand in hand with the establishment of general laws and theories.

Fundamental measurement based on stipulations of the kind discussed so far, while probably the only type of nonderivative measurement used in physics, is not the only type that is conceivable. Psychology, for example, has developed procedures of a quite different character for the fundamental measurement of various characteristics. As an example,[73] let us briefly consider a method, developed by S. S. Stevens and his associates,[74] for the measurement of pitch, which is an attribute of tones and must be distinguished from the frequency of the corresponding vibrations in a physical medium. The scale in question is obtained by means of "fractionation" experiments, in which a subject is presented with a pair of pure tones of fixed frequencies and then divides the pitch interval determined by them into what he per-

Derived Measurement

ceives as four phenomenally equal parts, with the help of an apparatus which permits the production of pure tones of any frequency between the two given ones. This type of experiment was performed repeatedly with several subjects and for different pairs of tones whose frequency ranges partly overlapped. The responses obtained from different observers were found to be in satisfactory agreement, and for different frequency intervals they dovetailed so as to determine one uniform scale. A certain pure tone was then chosen as standard and was assigned the arbitrary value of 1,000 mels as a measure of its pitch; and another pure tone, which marks the limit of pitch perception, was assigned zero pitch. Through these stipulations it was possible to assign, in a reasonably univocal fashion, numerical pitch values to the pure tones with frequencies of between 20 and 12,000.

The procedure just described differs plainly from the method analyzed before; in particular, it nowhere relies on any mode of combination. On the other hand, it presupposes no other scale of measurement (theoretically not even that of frequency); it has therefore to be qualified as fundamental measurement.[75]

There are indications that the measure of pitch thus determined may have theoretical import: It correlates with the scales obtained through somewhat different procedures (such as having the subject produce a tone which sounds "half as high" as a given other tone); and—what is more significant—there is evidence which suggests the existence of a physiological counterpart of the pitch scale: The distance in mels between pure tones seems to be proportional to the linear distances between those regions on the basilar membrane in the inner ear which are made to vibrate most vigorously by those tones; a relationship which does not obtain between the distances in frequency between pure tones and the corresponding basilar areas of maximal excitation.

13. Derived Measurement

By derived measurement we understand the determination of a metrical scale by means of criteria which presuppose at least

Some Basic Types of Concept Formation in Science

one previous scale of measurement. It will prove helpful to distinguish between derived measurement by stipulation and derived measurement by law.

The former consists in defining a "new" quantity by means of others, which are already available; it is illustrated by the definition of the average speed of a point during a certain period of time as the quotient of the distance covered and the length of the period of time. Derived measurement by law, on the other hand, does not introduce a "new" quantity but rather an alternative way of measuring one that has been previously introduced. This is accomplished by the discovery of some law which represents the magnitude in question as a mathematical function of other quantities, for which methods of measurement have likewise been laid down previously. Thus, certain laws of physics make it possible to use sound or radar echoes for measuring spatial distances by the measurement of time lapses. Other instances are the measurement of altitude by barometer, of temperature by means of a thermocouple, and of specific gravity by hydrometer, as well as the trigonometric methods used in astronomy and other disciplines in determining the distances of inaccessible points as functions of other distances which are amenable to more direct measurement; these latter methods are based, in particular, on the laws of physical geometry.

While fundamental measurement gives rise to rational values only, its combination, in derived measurement, with general laws or theories, calls for the admission of irrational values as well. Thus, e.g., when direct measurement has yielded the length 10 for the sides of a square, geometry demands that its diagonal be assigned the irrational number $\sqrt{200}$ as its length, although fundamental measurement could never establish or disprove this assignment.[76] Similarly, the law that the period t of a mathematical pendulum is related to its length l by the formula $t = 2\pi\sqrt{l/g}$ requires irrational and even transcendental values for the periods of some pendulums, although again fundamental measurement cannot prove or disprove such an assignment.

As we have seen, the rules of fundamental measurement determine the value of a quantity only for objects of a certain in-

Derived Measurement

termediate range, to which we referred as constituting a domain D. In the fundamental measurement of mass, for instance, D consists of those physical bodies which are capable of being weighed on a balance; in the fundamental measurement of length, D is the class of all physical distances capable of direct measurement by means of yardsticks. But the use of such terms as 'mass' and 'length' in physical theory extends far beyond this domain: physics ascribes a mass to the sun and a length to the distance between the solar system and the Andromeda nebula; it determines the masses of submicroscopic particles and the wave lengths of X-rays; and none of these values is obtainable by fundamental measurement.

Thus, also for this reason, the rules for the fundamental measurement of a quantity s do not completely define s, i.e., they do not determine the value of s for every possible case of its theoretically meaningful application. To give an interpretation to 's' outside the original domain D, an extension of those rules is called for. The same remark applies to many quantities for which rules of indirect measurement have been laid down, such as, say, the concept of temperature as interpreted by reference to mercury as a thermometric substance.

One important type of procedure for extending the rules of measurement for a given quantity s consists in combining the methods of derived measurement by law and by stipulation. Suppose, for example, that a scale for the measurement of temperature is originally determined (through derived measurement by stipulation) by reference to a mercury thermometer. Then the concept of temperature is interpreted only for substances which fall within the range between the melting point and the freezing point of mercury. Now, it is an empirical law that, within this range, the temperature of a body of gas under constant pressure can be represented as a specific mathematical function, f, of its volume: $T = f(v)$. This law provides a possibility for the indirect measurement of temperature of a substance by means of a "gas thermometer," i.e., by determining the volume v which a certain standard body of gas assumes under a specified pressure when brought in contact with that substance;

Some Basic Types of Concept Formation in Science

the temperature will then be $f(v)$. Evidently, the formula '$T = f(v)$' represents an empirical law only within the range of the mercury thermometer, since, for other cases, 'T' has received no interpretation. But it is possible to extend its domain of validity by fiat, namely, by stipulating that, outside the range of the mercury thermometer, the formula '$T = f(v)$' is to serve as a definition—or, rather, as a partial interpretation—of the concept of temperature; i.e., that the temperature of a substance outside the range of the mercury thermometer is to be set equal to $f(v)$, where v is the volume which a certain standard body of gas assumes when brought in contact with the substance. Thus, the gas thermometer can now be used for the derived measurement of temperature in a much larger domain than that covered by the mercury thermometer; within the range of the latter, the use of the gas thermometer represents indirect measurement by law; outside that range, indirect measurement by stipulation.

The same type of procedure is used for further extensions of the temperature scale and for the extension of other metrical scales as well. Thus, e.g., the use of trigonometric methods for the determination of certain astronomical distances and the reliance on gravitational phenomena in determining the masses of astronomical bodies may be viewed as extensions of the rules for the fundamental measurement of length and of mass and as providing indirect measurement by law within the original range of the latter and by stipulation outside that range.

It should be noted, however, that this brief account of the interpretation of metrical terms in science as a process of piecemeal stipulative extension of low-order empirical generalizations is considerably schematized in order to exhibit clearly the basic structure of the process. In practice, the empirical "laws" (such as '$T = f(v)$' above) which form the basis of the process often hold only approximately, and there may be considerable deviations from it, particularly at the ends of the original scale. In such cases the original scale of measurement for the given magnitude may be dropped altogether in favor of the more comprehensive scale of indirect measurement; thus, e.g., the temperature scale determined by the gas thermometer approximately, but not strictly, coincides in the range of the mercury thermometer with the scale determined by the latter; it was therefore used to replace the latter in the interest of securing an interpretation of the concept of temperature which would be unambiguous and which would cover a wider range of cases. The process did not stop there, however. For the purposes of theoretical physics, a thermodynamical scale of temperature was eventually introduced which permitted, in conjunction with other concepts, the formulation of a system of thermodynamics distinguished by its theoretical power and its formal simplicity. The thermodynamic concept of temperature has the status of a theoretical construct; it is introduced, not by reference to any particular thermometric substance, but by laying down, hypothetically, a set of general laws couched in terms of it and some other constructs and by providing a partial empirical interpretation for it or for certain derivative terms. The various methods for the measurement of temperature have to be viewed as partial and approximate interpretations of this theoretical construct.

In a similar fashion considerations of theoretical import and systematic simplicity govern the gradual development of the rules for the measurement of many other quantities in the more advanced branches of empirical science; and frequently there is a complex interplay between the development of theoretical knowledge in a discipline and the criteria used for the interpretation of its metrical terms.

Thus, e.g., the concept of time, or of temporal duration of events, represents a theoretical construct whose empirical interpretation has undergone considerable changes. In principle any periodic process might be chosen for the determination of a time scale; the periodically repeated elementary phases of the process would be said to be of equal duration, and the temporal duration of a given event would be measured—to put it briefly—by determining the number of successive elementary processes which take place during the event. Thus, to give an example which Moritz Schlick used in his lectures, the pulse beat of the Dalai Lama might be chosen as the standard clock; but—apart from its enormous technical inconvenience—this convention would have the consequence that the speed of all physical processes would depend on the state of health of the Dalai Lama; thus, e.g., whenever the latter had a fever and showed what, by customary standards, is called a fast pulse, then such events as one rotation of the earth about its axis or the fall of a rock from a given height would take up more temporal units—and would therefore be said to take place more slowly—than when the Dalai Lama was in good health. This would establish remarkable laws connecting the state of health of the Dalai Lama with all events in the universe—and this by instantaneous action at a distance; but it would preclude the possibility of establishing any laws of the simplicity, scope, and degree of confirmation exhibited by Galileo's, Kepler's, and Newton's laws. The customary choice of the earth's daily rotation about its axis as a standard process recommends itself, among other things, because it does not have such strikingly undesirable consequences and because it does permit the formulation of a large body of comprehensive and relatively simple laws of physical change. Eventually, however, those very laws compel the abandonment of the daily rotation of the earth as the standard process for the measurement of time; for they entail the consequence that tidal friction and other factors slowly decelerate that rotation, so that the choice of the earth as a clock would make the speed of physical phenomena dependent upon the age of the earth and would thus have a similar effect as reliance on the pulse beat of the Dalai Lama. This consideration calls for the use of other types of standard processes, such as electrically induced vibrations of a quartz crystal.

Thus, the principles governing the measurement of time and temperature and similarly of all the other magnitudes referred to in physical theory represent complex and never definitive modifications of initial "operational" criteria; modifications which are determined by the objective of obtaining a theoretical system that is formally simple and has great predictive and explanatory power: Here, as elsewhere in empirical science, concept formation and theory formation go hand in hand.[77]

14. Additivity and Extensiveness

A distinction is frequently made between additive and nonadditive quantities and similarly between extensive and intensive characteristics.[78] The ways in which different authors construe these concept pairs conflict to some extent, and some of the criteria offered to explicate them involve certain difficulties. By reference to our preceding analyses, we will now restate concisely what appears to be the theoretical core of those ideas.

The distinction of additive and nonadditive quantities refers to the existence or nonexistence, for a given quantitative concept, of an operational interpretation for the numerical addition of the s-values of two different objects. In this sense, length is called an additive quantity because the sum of two numerical length-values can be represented as the length of the interval obtained by joining two intervals of the given lengths end to end in a straight line; temperature is said to be nonadditive because there is no operation on two bodies of given temperatures which will produce an object whose temperature equals the sum of the latter. To state this idea more precisely, we first define a relative concept of additivity:

(14.1) A quantity s is additive relatively to a combining operation o if $s(x \circ y) = s(x) + s(y)$ whenever x, y, and $x \circ y$ belong to the domain within which s is defined

A quantity may be additive relatively to some mode of combination, nonadditive relatively to others. Thus, e.g., the electric resistance of wires is additive relatively to their arrangement in series, nonadditive relatively to their arrangement in parallel. The reverse holds for the capacitance of condensers. Again, the length of intervals marked off on metal rods is additive relatively to the operation of placing them horizontally end to end along a straight line but not strictly additive if the rods are placed on one another vertically in a straight line; in this case the length of the combination will be somewhat less than the sum of the lengths of the components. However, in prevailing usage the terms 'additive' and 'nonadditive' are not relativized with regard to some specified mode of combination: rather, they serve to qualify given quantities categorically as additive or nonadditive. Can such usage be given a satisfactory explication? Can we not simply call a quantity additive if there exists *some* mode of combination in regard to which the quantity satisfies (14.1)? No; this criterion would classify as additive many quantities which in general usage would be called nonadditive (strictly speaking, any quantity is additive under this rule); thus, e.g., the temperature of gases would be additive, for the conditions (14.1) are satisfied by the operation of mixing two bodies of gas and then heating the mixture until its temperature equals the sum of the initial temperatures of its components. To do justice to the intent of the notion of additivity, we would have to preclude such complicated and "artificial" modes of combination as this and to insist on simple and "natural" ones.[79] In the light of our earlier discussions, these two qualifications have to be understood, not in the psychological sense of intuitive simplicity and familiarity, but in the systematic sense of theoretical simplicity and fruitfulness. This suggests the following explication:

(14.2) A quantity s is additive if there exists a mode of combination o such that (1) s is additive relatively to o in the sense of (14.1); (2) o together with s gives rise to a simple and fruitful theory

We now turn to the distinction of extensive and intensive characteristics. It is intended to divide all attributes permitting distinctions as to more or less into two groups: those which can and those which cannot be metricized by the basic physical method of fundamental measurement. Now, any attribute capable of gradations determines a comparative concept in the sense of section 11, and its amenability to fundamental measurement requires the existence of a mode of combination in regard to which the conditions of extensiveness are satisfied. We therefore define:

(14.3) A comparative concept represented by two relations, C and P, which determine a quasi-series within a class D is

Additivity and Extensiveness

extensive relative to a given mode of combination, o, if the conditions (12.4) are all satisfied

But, again, the term 'extensive' is customarily used in a non-relativized form. Could we interpret this usage by calling a comparative concept extensive if there exists *some* mode of combination relative to which the concept is extensive in the sense of (14.3)? No; just as in the case of additivity, this criterion is too weak; for it can be shown that if a comparative concept can at all be metricized in the sense of (12.1)—no matter whether derivatively or by one or another kind of fundamental measurement—then there exists some mode of combination, albeit possibly a rather "artificial" one, which satisfies (14.3); hence any comparative concept of this sort would have to count as extensive, which is quite contrary to customary usage. Considerations analogous to those that led to (14.2) suggest the following explication of the distinction between extensive and intensive characteristics:

(14.4) A comparative concept represented by two relations, C and P, which determine a quasi-series within a class D is extensive (intensive) if there exists some (no) mode of combination o such that (1) C, P, and o jointly satisfy the conditions (12.4) of extensiveness; (2) o together with C and P gives rise to a simple and fruitful theory

In the sense of this explication, the comparative concept of mass characterized by the stipulations (11.3) is extensive in its domain of application, since there exists a "simple" and "natural" mode of combination, namely, the operation j specified in (12.2b), in regard to which the conditions of extensiveness are satisfied. On the other hand, consider the comparative concept of hardness determined by the scratch test, which establishes the ordering relations "as hard as" and "less hard than." No simple and natural mode of combining minerals is known which, jointly with those two relations, satisfies the conditions of extensiveness. (The joining operation j, for example, will not do at all; for the join of two different mineral objects is, as a rule, an inhomogeneous body, to which the scratch test is not applicable.) Hardness as characterized by the scratch test has to be qualified, therefore, as an intensive characteristic.

Obviously, an intensive characteristic as here construed is not capable of fundamental measurement by reference to some mode of combination which is governed by simple theoretical principles; but it may well be amenable to some alternative type of fundamental measurement (as are tonal pitch and many other objects of psychophysical measurement) or to derivative measurement (as are density, temperature, refractive index, etc.).

If the explications which have been propounded in this section do justice to the theoretical intent of the two concept pairs under analysis, then a few simple consequences follow, which we now state in conclusion:

Whether a given quantity is qualified as additive or non-additive and, similarly, whether an attribute is adjudged extensive or intensive will depend on the theoretical knowledge available. In classical mechanics, for example, mass and the speed of rectilinear motion are additive; in relativistic physics, they are not.

Furthermore, there are no sharp boundary lines separating extensive from intensive and additive from nonadditive concepts. For the criteria on which these distinctions are based invoke the concepts of theoretical simplicity and fruitfulness, both of which are surely capable of gradations.

And, lastly, the very fact that questions of simplicity and systematic import enter into the criteria for the distinction reflects once more the pervasive concern of empirical science: to develop a system of concepts which combines empirical import with theoretical significance.

Notes

Preliminary remark. Abbreviated titles inclosed in brackets refer to the Bibliography at the end. Several of the previously published monographs of the *International Encyclopedia of Unified Science* contain material relevant to the problems of concept formation. For convenience, those monographs will be referred to by abbreviations; 'EI3', for example, indicates Volume I, No. 3.

1. I am gratefully indebted to the John Simon Guggenheim Memorial Foundation, which granted me, for the academic year 1947–48, a fellowship for work on the logic and methodology of scientific concept formation. The present monograph is part of the outcome of that work. I sincerely thank all those who have helped me with critical comments or constructive suggestions; among them, I want to mention especially Professors Rudolf Carnap, Herbert Feigl, Nelson Goodman, and Ernest Nagel, Dr. John C. Cooley, and Mr. Herbert Bohnert.

2. Thus, e.g., the genus-and-differentia rule is explicitly advocated in Hart [Report], which presents the views of a special committee on conceptual integration in the social sciences.

3. Quine [Math. Logic], p. 47.

4. Hutchinson [Biology]. (Quoted, with permission of the editor, from the 1948 copyright of Encyclopaedia Britannica.)

5. The concepts of determinacy and uniformity of usage as well as those of vagueness and inconsistency of usage are relative to some class of individuals using the language in question; they are therefore pragmatic rather than syntactic or semantic in character. On the nature of pragmatics, semantics, and syntax see Carnap [EI3], secs. 1, 2, and 3, and Morris [EI2].

6. See [Log. Found. Prob.], chap. i.

7. For details cf. Russell [Math. Philos.] and Tarski [Truth].

8. Cf., for example, Sommerhoff [Analyt. Biol.], which combines a lucid presentation of the general idea of explication with some useful object lessons in the explication of certain fundamental concepts of biology.

9. A spirited discussion of this issue by a group of psychologists and social scientists may be found in Sargent and Smith [Cult. and Pers.], esp. pp. 31–55. This debate illustrates the importance, for theorizing in psychology and the social sciences, of a clear distinction between the various meanings of "(real) definition."

10. Explicit (though not the fullest possible) use of this mode of formulation is made by Hogben in the presentation of his auxiliary international language, Interglossa. His English translations of Interglossa phrases include such items as these:

habe credito ex Y = owe Y; date credito Y de Z = lend Y (some) Z; X acte A Y = X performs the action A on Y ([Interglossal], pp. 45 and 49).

Similarly, Lasswell and Kaplan, in [Power and Soc.], use variables to indicate the syntax of some of their technical terms. Thus, e.g., power is defined as a triadic relation: "*Power* is participation in the making of decisions: G has power over H with respect to the values K if G participates in the making of decisions affecting the K-policies of H" (*ibid.*, p. 75). Note that this definition is expressed contextually rather than by simulating the genus-and-differentia form, which is strictly inapplicable here.

Notes

11. Definitions of the form considered in the present and the preceding sections are often called *explicit definitions*. They state explicitly, in the definiens, an expression which is synonymous with the expression to be defined and in favor of which the latter can always be eliminated. They differ in this respect from the so-called *recursive definitions*, which play an important role in logic and mathematics but are not used in empirical science. For details on the formal aspects of explicit and recursive definition cf. Church [Articles] and Carnap [Syntax]. A critical historical study of various conceptions of definition, together with an examination of the function of definition in mathematics and empirical science, may be found in Dubislav [Definition]. Lewis [Analysis] contains a detailed discussion of definition, with special emphasis on explication. Various important observations on the nature and function of definition are included in Quine [Convention]. Chapter i of Goodman [Appearance] is an excellent study of the use of definition in rational reconstruction. Robinson [Definition] presents a nontechnical discussion of various aspects of definition; this book is not predominantly concerned, however, with definition in science.

12. For a fuller statement, and a partly critical discussion, of this injunction see Eaton [Logic], chap. vii.

13. Hilbert [Grundlagen], § 3.

14. Peano [Définitions].

15. Important contributions to the clarification and partial solution of this problem are contained in Carnap [Aufbau] and in Goodman [Appearance].

16. Cf. [Forschung], secs. 25–30.

17. See, e.g., Woodrow [Laws]; Hull [Int. Var.]; Spence [Theory Construction]; and MacCorquodale and Meehl [Distinction].

18. These were first pointed out by Carnap in [Testability].

19. For details on this problem and its ramifications see Chisholm [Conditional]; Goodman [Counterfactuals]; Hempel and Oppenheim [Explan.]; Lewis [Analysis], chap. vii; Reichenbach [Logic], chap. viii.

20. In [Testability]; a less technical account may be found in Parts III and IV of Carnap's [EI1].

21. It might be objected that these remarks disregard the distinction between (1) *assigning a meaning* to a scientific concept by reference to observables and (2) *discovering an empirical regularity* which connects the previously defined concept with certain observables. Such discovery, it might be argued, though providing new criteria of application for the concept, does not affect its meaning at all. Now, I think it is often useful to make a distinction between questions of meaning and questions of fact; but I have doubts about the possibility of finding precise criteria which would explicate the distinction. For this reason, the presumptive objection does not seem to me decisive. Reasons for this view may be found in White [Analytic] and Quine [Dogmas]. This issue, however, is still the object of considerable controversy, and my remarks in the text are therefore deliberately sketchy.

22. This is, in effect, the liberalized version of physicalism by which Carnap proposed to replace the earlier form, to which we referred as the narrower thesis of empiricism. More specifically, the liberalized thesis asserts that each term of empirical science can be introduced by means of an introductive chain, i.e., an ordered set of reduction sentences analogous to a definition chain; for details see Carnap [Testability], esp. secs. 9, 15, and 16; a less technical synopsis is given in Carnap [EI1], Secs. III and IV.

Notes

23. More precisely, an introductive chain as defined by Carnap can be shown to specify, in terms of observables, one necessary and one sufficient condition for the term it introduces. The bilateral reduction sentence (6.3), e.g., which is a simple special case of an introductive chain, specifies that $P_1x \cdot P_2x$ is a sufficient condition for Qx, and $\sim(P_1x \cdot \sim P_2x)$ is a necessary one. These conditions apply quite generally to any object x, no matter whether it meets the test condition P_1x or not. Carnap has not discussed in detail reduction sentences for expressions involving more than one variable; but these are called for by the liberalized physicalistic thesis, and his theory of reduction can readily be transferred to this case. But, in doing so, we obtain the consequence that an introductive chain for 'length $(u, v) = r$' must specify a necessary and a sufficient condition for every sentence of this form, i.e., for every possible set of values r (> 0), u, v. And this clearly cannot be accomplished in terms of observation predicates.

24. For fuller details on the axiomatic method cf., e.g., Tarski [Logic], chaps. vi ff., and Woodger's treatises [Ax. Meth.] and [EII5]. It should be noted that the conception of scientific theories as presented in axiomatized form is an idealization made for purposes of logical clarification and rational reconstruction. Actual attempts to axiomatize theories of empirical science have so far been rare. Apart from different axiomatizations of geometry, the major instances of such efforts include Reichenbach's [Axiomatik], Walker's [Foundations], the work of Woodger just referred to, the axiomatization of the theory of rote learning by Hull and his collaborators (cf. [Rote Learning]), and, in economic theory, axiomatic treatments of such concepts as utility (see, e.g., the axiomatization of utility in von Neumann and Morgenstern [Games], chap. iii and Appendix).

25. See [Grundlagen].

26. It is sometimes held that in an uninterpreted axiomatized theory the axioms, or postulates, themselves constitute "implicit definitions" of the primitives and that, accordingly, the latter mean just what the postulates require them to mean. This view has been eloquently advocated by Schlick (cf. [Erkenntnislehre], sec. 7) and by Reichenbach (see [Raum-Zeit-Lehre], § 14), and, more recently, it has been invoked by Northrop (see, e.g., [Logic], chap. v), and it is reflected also in Margenau's concept of a "constitutive definition" for theoretical constructs (cf. [Reality], chap. xii). This conception of the function of postulates faces certain difficulties, however. According to it, the term 'point' in pure Euclidean geometry means an entity of such a kind that, for any two of them, there exists exactly one straight line on which both are incident, etc. But since the terms 'straight line', 'incident', etc., have no prior meaning assigned to them, this characterization cannot confer any specific meaning on the term 'point'. To put it differently: The conjunction of all the postulates of an axiomatized theory may be construed as a sentential function in which the primitives play the role of variables. But a sentential function cannot well be said to "define" one particular meaning of (i.e., one particular set of values for) the variables it contains unless proof is forthcoming that there exists exactly one set of values for the variables which satisfies the given sentential function. The postulates of an uninterpreted deductive system may well be said, however, to impose limitations upon the possible interpretations for the primitives. Thus, e.g., the postulate that for two points there exists exactly one straight line on which both are incident precludes the interpretation of 'point' by 'person', 'line' by 'club', and 'incident on' by 'member of', for this interpretation would turn the postulate into a false sentence.

For a concise discussion of the notion of implicit definition and its historical roots see

Notes

Dubislav [Definition], secs. 28 and 29; on the origin of the idea in Gergonne's work see also Nagel [Geometry], secs. 27-30.

27. For details on pure and physical geometry, and for further references, see Carnap [EI3], secs. 21 and 22; Hempel [Geometry]; and especially Reichenbach [Rise], chap. viii, and the detailed work [Raum-Zeit-Lehre]. On the relativistic treatment of physical geometry, cf. also Finlay-Freundlich [EI8].

28. A distinction between abstract theoretical system and interpretation is made by Campbell, who divides a physical theory into the "hypothesis," i.e., a set of propositions "about some collection of ideas which are characteristic of the theory," and the "dictionary"; the latter relates to some (but not to all) propositions of the hypothesis certain empirical propositions whose truth or falsity can be ascertained independently, and it states that one of these sets of related propositions is true if and only if the other is true ([Physics], p. 122). Note that the argument presented earlier in section 7 casts doubt upon the "if and only if." In a similar vein, Reichenbach speaks of the "coördinative definitions" which, by coördinating physical objects with geometrical concepts, specify the denotations of the latter (cf. [Rise], chap. viii; [Raum-Zeit-Lehre], sec. 4). Carnap, in [EI3], secs. 21–24, analyzes the empirical interpretation of abstract calculi as a semantical procedure. The logical structure of scientific theories and their interpretation has recently been discussed at length also by Northrop ([Logic], esp. chaps. iv–vii, and [Einstein]) and by Margenau ([Reality], chaps. iv, v, and xii); the rules effecting the interpretation are called "epistemic correlations" by Northrop, "rules of correspondence" by Margenau. (Both authors envisage an interpretation in phenomenalistic terms rather than by reference to intersubjectively ascertainable observables.) Einstein's lecture, [Method], contains a lucid discussion of the problem at hand in special reference to theoretical physics.

29. See [Ax. Meth.] and [EII5].

30. Two of the monographs in the *International Encyclopedia of Unified Science* give brief accounts of the fundamental ideas of the kinetic theory: Lenzen [EI5], sec. 15, and Frank [EI7], Part IV. For concise analyses of the hypotheses and interpretations involved in this case see Campbell [Physics], pp. 126–29, and Nagel [Reduction], pp. 104–11.

31. Koch's essay, [Motivation], offers some good examples of the interpretation of theoretical constructs in psychology by means of empirical terms which, in turn, are introduced by chains of reduction sentences based on observation terms.

32. Thus, e.g., A. Wald says on the interpretation of a scientific theory, "In order to apply the theory to real phenomena, we need some rules for establishing the correspondence between the idealized objects of the theory and those of the real world. These rules will always be somewhat vague and can never form a part of the theory itself" ([Statist. Inf.], p. 1).

33. Cf. Kaplan [Def. and Spec.].

34. Cf. Carnap [Log. Found. Prob.], chap. ii; Nagel [EI6]; Reichenbach [Probability], chaps. ix and xi; also see Helmer and Oppenheim [Degree], where a theory of logical probability is developed which differs from Carnap's in various respects.

35. Cf. especially [Modern Physics], [Physical Theory], [Op. An.], and [Concepts]. For a concise and lucid presentation and appraisal of the central ideas of operationism see Feigl [Operationism]. For enlightening comments on operationism in psychology, cf. Bergmann and Spence [Operationism] and Brunswik [EI10], chaps. i and ii.

Notes

36. [Op. An.], p. 119. In this article, incidentally, Bridgman points out: "I believe that I myself have never talked of 'operationalism' or 'operationism', but I have a distaste for these grandiloquent words which imply something more philosophic and esoteric than the simple thing that I see" (*ibid.*, p. 114).

37. Cf. Bridgman [Op. An.] and [Concepts].

38. For elementary accounts see Ayer [Language], chap. i and Preface; Pap [Anal. Philos.], chap. xiii. A more advanced treatment may be found in Carnap [Testability]; recent critical surveys in Hempel [Emp. Crit.] and [Cogn. Signif.].

39. [Institutions].

40. In some cases the reliability of a test is intended to be an index of the objective consistency of its different components. For a fuller discussion see Thurstone [Reliability] and Guilford [Methods]. Dodd's article [Op. Def.] contains useful comments on the notion of reliability.

41. Cf. Ogburn [Social Change], Parts IV and V; Ogburn and Nimkoff [Sociology], pp. 881 ff.

42. [Social Change], p. 297.

43. The discussion of cultural lag in Ogburn and Nimkoff [Sociology], pp. 881 ff., takes cognizance of this difficulty and proposes certain qualified formulations; these, however, still contain valuational terms such as "best adjustment to a new culture trait," "harmonious integration of the parts of culture," etc., so that the basic objection raised here remains unaffected.

44. To suggest a direction in which such a restatement might be sought, let us note that lag phenomena are known in the area of the physical sciences as well. Thus, when a jar containing a viscous liquid is tilted, a certain period of time elapses before the liquid has "adjusted" itself to the new position of the jar; again, the changes in a magnetic field and the resultant magnetization of a piece of steel in the field exhibit a temporal lag in that the maxima and minima of the latter "lag behind" those in the intensity of the field; however, a theoretical account of the first phenomenon uses descriptive concepts such as that of equilibrium rather than the normative concept of good adjustment; and, in the second case, the theoretical analysis does not assert that the maxima and minima of induced magnetization "should" occur earlier than they do but simply aims at a description of the temporal lag by means of graphs or mathematical functions. It is conceivable that the empirical information which the hypothesis of cultural lag is intended to convey might be satisfactorily restated in an analogous fashion. On this point see also Lundberg's discussion of cultural lag and related concepts in [Foundations], pp. 521 ff.

45. Cf., e.g., Malinowski [Dynamics] and the searching study of functional analysis presented in Merton [Social Theory], chap. i.

46. A rigorous explication of the notions of scope, confirmation, and formal simplicity presents considerable difficulties; but, for our present purposes, an intuitive understanding of these concepts will suffice. The characteristics of a good theory are concisely surveyed in Nagel [EI6], sec. 8. The concept of confirmation is dealt with in the publications listed in n. 34. Certain aspects of the intriguing notion of simplicity are examined in Popper [Forschung], secs. 31–46, and in Goodman [Appearance], chap. iii. Reichenbach [Experience], § 42, suggests a distinction between descriptive and inductive simplicity.

47. Cf. [Physique] and [Temperament].

Notes

48. For fuller details, cf. Hempel and Oppenheim [Typusbegriff], in Sheldon [Physique], chap. v, and Lazarsfeld and Barton [Qual. Meas.].

49. Cf., e.g., Dodd's "S-theory" (expounded most fully in [Dimensions]), which actually is not a sociological theory but a system of terminology and classification whose theoretical import is quite problematic.

50. For illustrations and a more detailed discussion of the concept of validity see Adams [Validity]; Dodd [Op. Def.]; Guilford [Methods], pp. 279 and 421; Thurstone Reliability], sec. 25.

51. The second criterion is applied, e.g., by Stevens and Volkmann (cf. [Pitch]), who claim validity for their pitch scale partly on the ground that it permits fitting into one simple curve the data obtained by several sets of experiments, and who also mention a law of simple mathematical form that appears to connect pitch with a certain anatomical aspect of the hearing process (cf. pp. 68 and 69 of this monograph).

52. This is stressed by Lundberg in his critique of common usage as a standard for the operational interpretation of sociological terms; see [Definitions] and, especially, [Measurement].

53. An entirely different view is expressed by G. W. Allport, who insists that "mathematical or artificial symbols" are not suited to name human traits and that "the attributes of human personality can be depicted only with the aid of common speech, for it alone possesses the requisite flexibility, subtlety, and established intelligibility" ([Personality], p. 340). In a similar vein, Allport and Odbert assert, concerning the naming of traits, "Mathematical symbols cannot be used, for they are utterly foreign to the vital functions with which the psychologist is dealing. Only verbal symbols (ambiguous and troublesome as they are) seem appropriate" ([Trait-Names], p. v). In accordance with this view, the authors hold that "the empirical discovery of traits in individual lives is one problem, that of selecting the most appropriate names for the traits thus discovered is another" (*ibid.*, p. 17).

54. Cf. L. L. Thurstone [Vectors], [Analysis], and [Abilities]; for detailed references to the comprehensive literature see Wolfle [Factor Analysis].

55. At present, the theoretical significance of the primary characteristics evolved by factor analysis appears to reside mainly in their ability to permit a formally simplified, or economical, descriptive representation of human traits; and, indeed, Thurstone repeatedly emphasizes considerations of theoretical parsimony (cf., e.g., [Vectors], pp. 47, 48, 73, 150, 151; [Analysis], p. 333). The simplicity thus achieved is of the kind which Reichenbach calls descriptive and which he distinguishes from inductive simplicity; the latter is closely related to predictive power and thus to theoretical import (cf. [Experience], sec. 42). The predictive aspect is not strongly emphasized in Thurstone's work (cf., e.g., [Analysis], pp. 59 ff.); but there are certain remarks as to the possibility of a genetic substructure underlying the system of factors, which would confer upon factor analysis special significance for the study of mental inheritance and of mental growth (cf. Thurstone [Vectors], pp. 51 and 207; and [Analysis], p. 334). If such connections could be ascertained, they would provide the conceptual systems developed by factor analysis with theoretical import in addition to the systematic advantage of descriptive simplicity. At present, however, few if any suggestions of general laws (of causal or statistical form) can be found in the work on factor analysis—except for the statement of independence of certain primary traits. See also Brunswik's discussion, in [EI10], of the issue at hand.

56. Cf. the discussion in Spence [Learning].

Notes

57. Mayr [Systematics], p. 10.

58. Huxley [New Syst.], p. 20. Fuller details on this conception of natural classification in modern taxonomy may be found in Gilmour [Taxonomy], Huxley, *op. cit.*, and Mayr [Systematics].

59. On the methodology of classificatory and related procedures in the social sciences see Lazarsfeld and Barton [Qual. Meas.].

60. This terminology is Carnap's; see [Log. Found. Prob.], secs. 4 and 5, where he distinguishes and compares classificatory, comparative, and quantitative concepts.

61. Not a numeral (i.e., a symbol naming a number), as has been asserted by several authors, including Campbell (cf. [Physics], chap. x; [Measurement], p. 1 *et passim;* and Campbell's contribution to Ferguson [Reports]), Reese ([Measurement]), and Stevens (see, e.g., [Scales] and [Math., Meas., and Ps.]). In [Physics], chap. x, p. 267, however, Campbell declares: "Measurement is the process of assigning numbers to represent qualities." And, indeed, the values of quantitative concepts have to be construed so as to be able to enter into mathematical relationships with each other, such as those expressed by Newton's law of gravitation, the laws for the mathematical pendulum, Boyle's law, etc. They must, therefore, permit multiplication, the extraction of roots, etc.; and all these operations apply to numbers, not to numerals. Similarly, it is impossible to speak significantly of the distance, or difference, of two numerals.

62. This survey follows in part Carnap's discussion in [Begriffsbildung], pp. 51–53.

63. Properly speaking, a balance compares and measures gravitational forces exerted by the earth upon bodies placed in the scales. Since these forces, however, are proportional to the masses of those objects, an "operational definition" of mass can be based on the use of a scale. In a similar vein, Frank ([E17], p. 12) gives an "operational definition" of the mass of a body as the number of grams ascertained by reading its weight on a spring balance at sea level. For a fuller discussion of the point at hand see Campbell [Measurement], chap. iii, and esp. p. 45.

64. For further details on the problems of ordering and measurement discussed in the following sections cf. Campbell's analyses in [Physics], Part II, and in [Measurement]; Carnap [Begriffsbildung]; von Helmholtz [Zählen und Messen]: Hempel and Oppenheim [Typusbegriff] (on comparative concepts); Lenzen [EI5] (esp. secs. 5, 6, and 7, which deal with the measurement of length, time, and weight); Nagel [Measurement] and [Log. of Meas.]; Russell]Principles], Part III; Stevens [Math., Meas., and Ps.]; and Suppes' precise study [Ext. Quant.].

65. In logic, a relation P is said to constitute a series within a class D if within D it is transitive, irreflexive (i.e., for no x in D does xPx hold), and connected (i.e., if not $x = y$ then xPy or yPx). Our concepts of C-irreflexivity, C-connectedness, and quasi-series are generalizations of these ideas and include the latter as the special case where C is the relation of identity. Let us note that, among others, the following theorems are consequences of the conditions specified in (11.2):

(T1) $\qquad (x)(y) \, \{xCy \equiv (u) \, [(xPu \equiv yPu) \cdot (uPx \equiv uPy)]\}$
(T2) $\qquad (x)(y)(z) \, [(xCy \cdot yPz) \supset xPz]$

(T1) corresponds closely to a definition of coincidence in terms of precedence as used by Campbell; cf. [Measurement], p. 5.

66. Cf. the discussion of this point in Campbell [Measurement], chap. iii.

67. Cf. Bollenrath [Härte], where an account of more recent alternative comparative and metrical concepts of hardness may also be found.

Notes

68. See [Physique] and [Temperament].

69. This account of the fundamental measurement of mass is necessarily schematized with a view to exhibiting the basic logical structure of the process. We have to disregard such considerations as that the equilibrium of a balance carrying a load in each pan may not be disturbed by placing into one of the pans an additional object which is relatively light but whose mass is ascertainable by fundamental measurement. This means that fundamental measurement does not assign exactly one number to every object in D_1. A more detailed discussion of problems of this type may be found in Campbell [Measurement].

70. This "hence" relies on yet another physical fact, namely, that fundamental measurement assigns to any object in D_1 a positive number. This would not be the case but for the stipulation, included in our definitions of balancing and outweighing, that the balance be placed in a vacuum; for, in air, objects such as helium-filled balloons would have to be assigned negative numbers. See Campbell's remarks on this point in [Physics], pp. 319–20, and [Measurement], pp. 37–38.

71. A complete statement of the conditions here referred to does not seem to exist in the literature. The most detailed analysis appears to be Campbell's in [Measurement]. Other important contributions to the problem may be found in Helmholtz [Zählen und Messen]; Hölder [Quantität]; Nagel [Log. of Meas.] and [Measurement]; Suppes [Ext. Quant.]. Concerning the conditions of extensiveness, let us note here that we cannot require 'xoy' to have a meaning for every x and y in D. Indeed, the modes of combination on which fundamental measurement is based are usually inapplicable when x and y are identical or have common parts. (This is clearly illustrated by the operation j we invoked in the fundamental measurement of mass.) Nor can we require that D be closed under the operation o, i.e., that whenever x and y belong to D then xoy belongs to D again; for—to illustrate by reference to the operation j—the joining of two large bodies each of which can just barely be weighed on a scale will yield an object to which this procedure is no longer applicable.

The conditions laid down under (12.4) should accordingly be construed as applying to all those cases where the combinations mentioned exist and belong to D; and these cases are assumed to form a reasonably comprehensive set.

72. This conception was suggested to me by Professor Carnap.

73. For an account of other methods see Guilford [Methods]; Gulliksen [Paired Comparisons]; Thurstone [Methods].

74. Cf. Stevens and Volkmann [Pitch]; Stevens and Davis [Hearing], esp. chap. iii.

75. A similar method has been used by Stevens and his associates to establish scales of measurement for other attributes of tones; cf., e.g., Stevens [Loudness] and Stevens and Davis [Hearing]. Some authors, including Campbell, do not recognize this procedure as measurement, essentially on the ground that it is not of the extensive type, i.e., it does not rely on any mode of combination for the interpretation of addition. A critical discussion of Stevens' procedure from this point of view will be found in Ferguson [Reports], which includes a contribution by Campbell. For reasons mentioned in the present section, the conception of measurement which underlies this criticism appears unduly narrow; this is argued by Stevens in his article [Scales], which contains his direct reply, and in [Math., Meas., and Ps.]. For further light on this issue see Bergmann and Spence [Psych. Meas.] and Reese [Measurement].

76. For details on this important point, which again reflects the connection between theory formation and concept formation, and which also illustrates the grounds for re-

Notes

jecting the commensurability requirement (12.6), cf. von Helmholtz [Zählen und Messen] and Hertz [Comments], esp. p. 103; Campbell [Physics], pp. 310–13, and [Measurement], pp. 24 and 140; and Nagel [Op. An.], pp. 184–89.

77. The interplay of concept formation and theory formation is instructively exhibited in the "method of successive approximations," to whose importance in measurement Lenzen calls attention in [EI5]. This method for the successively more precise interpretation of metrical terms clearly presupposes the availability of certain laws and theories.

78. See, e.g., Cohen and Nagel [Logic], chap. xv, and Bergmann and Spence [Psych. Meas.]; also cf. Duhem [Théorie physique], pp. 177–80, where the basic idea underlying the two distinctions is vividly presented (though not logically analyzed) under the heading "Quantité et qualité."

79. In the context of their lucid analytic study, [Psych. Meas.], Bergmann and Spence have made an attempt to specify a restrictive clause for the admissible types of combination. They stipulate that the operation has to "lie within the dimension" of the quantity under consideration. This formulation is too elusive, however, to provide a solution to the problem. It might well be argued, for example, that our "artificial" operation, relatively to which the temperature of gases is additive, satisfies this requirement; for it involves measurement only in the "dimension" of temperature. For further observations on additivity, see especially Carnap [Begriffsbildung], pp. 32–35.

Bibliography

ADAMS, HENRY, F. [Validity] "Validity, Reliability, and Objectivity," in "Psychological Monographs," XLVII, No. 2 (1936), 329–50.

ALLPORT, GORDON W. [Personality] *Personality: A Psychological Interpretation.* New York, 1937.

ALLPORT, GORDON W., and ODBERT, HENRY S. [Trait-Names] *Trait-Names: A Psycho-lexical Study.* "Psychological Monographs," Vol. XLVII, No. 1 (1936).

AYER, ALFRED J. [Language] *Language, Truth and Logic.* 2d ed. London, 1946.

BERGMANN, GUSTAV, and SPENCE, KENNETH W. [Operationism] "Operationism and Theory in Psychology," *Psychological Review*, XLVIII (1941), 1–14.

———. [Psych. Meas.] "The Logic of Psychophysical Measurement," *Psychological Review*, LI (1944), 1–24.

BOLLENRATH, FRANZ [Härte] "Härte und Härteprüfung," in *Handwörterbuch der Naturwissenschaften*, Vol. V, ed. R. DITLER et al. 2d ed. Jena, 1931–35.

BRIDGMAN, P. W. [Modern Physics] *The Logic of Modern Physics.* New York, 1927.

———. [Physical Theory] *The Nature of Physical Theory.* Princeton, 1936.

———. [Op. An.] "Operational Analysis," *Philosophy of Science*, V (1938), 114–31.

———. [Concepts] "The Nature of Some of Our Physical Concepts," *British Journal for the Philosophy of Science*, I, 257–72 (1951); II, 25–44 and 142–60 (1951). Also reprinted as a separate monograph, New York, 1952.

BRUNSWIK, EGON. [EI10] *The Conceptual Framework of Psychology.* EI10. Chicago, 1952.

CAMPBELL, NORMAN R. [Physics] *Physics: The Elements.* Cambridge, England, 1920.

———. [Measurement] *An Account of the Principles of Measurement and Calculation.* London and New York, 1928.

CARNAP, RUDOLF. [Begriffsbildung] *Physikalische Begriffsbildung.* Karlsruhe, 1926.

———. [Aufbau] *Der logische Aufbau der Welt.* Berlin, 1928.

———. [Testability] "Testability and Meaning," *Philosophy of Science*, III (1936), 419–71, and IV (1937), 1–40. Also reprinted as a separate pamphlet (with corrigenda and additional bibliography by the author) by Graduate Philosophy Club, Yale University, 1950.

———. [Syntax] *Logical Syntax of Language.* London, 1937.

———. [EI1] "Logical Foundations of the Unity of Science," in EI1, pp. 42–62. Chicago, 1938. Reprinted in FEIGL and SELLARS [Readings].

———. [EI3] *Foundations of Logic and Mathematics.* EI3. Chicago, 1939.

———. [Log. Found. Prob.] *Logical Foundations of Probability.* Chicago, 1950.

Bibliography

CHAPIN, F. STUART. [Institutions] *Contemporary American Institutions.* New York, 1935.
CHISHOLM, RODERICK M. [Conditional] "The Contrary-to-Fact Conditional," *Mind,* LV (1946), 289–307. Reprinted in FEIGL and SELLARS [Readings].
CHURCH, ALONZO. [Articles] Articles "Definition" and "Recursion, Definition by," in D. D. RUNES (ed.), *The Dictionary of Philosophy.* New York, 1942.
COHEN, MORRIS R., and NAGEL, ERNEST. [Logic] *An Introduction to Logic and Scientific Method.* New York, 1934.
DODD, STUART C. [Dimensions] *Dimensions of Society.* New York, 1942.
———. [Op. Def.] "Operational Definitions Operationally Defined," *American Journal of Sociology,* XLVIII (1942–43), 482–89.
DUBISLAV, WALTER. [Definition] *Die Definition.* 3d ed. Leipzig, 1931.
DUHEM, P. [Théorie physique] *La théorie physique: Son objet et sa structure.* Paris, 1906.
EATON, R. M. [Logic] *General Logic.* New York, 1931.
EINSTEIN, ALBERT. [Method] *On the Method of Theoretical Physics.* Herbert Spencer Lecture. Oxford, 1933.
FEIGL, HERBERT. [Operationism] "Operationism and Scientific Method," *Psychological Review,* LII (1945), 250–59. Reprinted, with some alterations, in FEIGL and SELLARS [Readings].
FEIGL, HERBERT, and SELLARS, WILFRID (eds.). [Readings] *Readings in Philosophical Analysis.* New York, 1949.
FERGUSON, A. [Reports] FERGUSON, A., *et al.*, "Interim Report of Committee Appointed To Consider and Report upon the Possibility of Quantitative Estimates of Sensory Events," *Report of the British Association for the Advancement of Science, 1938,* pp. 277–334; and FERGUSON, A., *et al.*, "Final Report . . . ," *ibid., 1940,* pp. 331–49.
FINLAY-FREUNDLICH, E. [EI8] *Cosmology.* EI8. Chicago, 1951.
FRANK, PHILIPP. [EI7] *Foundations of Physics.* EI7. Chicago, 1946.
GILMOUR, J. S. L. [Taxonomy] "Taxonomy and Philosophy," in JULIAN HUXLEY (ed.), *The New Systematics,* pp. 461–74. Oxford, 1940.
GOODMAN, NELSON. [Counterfactuals] "The Problem of Counterfactual Conditionals," *Journal of Philosophy,* XLIV (1947), 113–28.
———. [Appearance] *The Structure of Appearance.* Cambridge, Mass., 1951.
GUILFORD, J. P. [Methods] *Psychometric Methods.* New York and London, 1936.
GULLIKSEN, H. [Paired Comparisons] "Paired Comparisons and the Logic of Measurement," *Psychological Review,* LIII (1946), 199–213.
HART, HORNELL. [Report] "Some Methods for Improving Sociological Definitions: Abridged Report of the Subcommittee on Definition of Definition of the Committee on Conceptual Integration," *American Sociological Review,* VIII (1943), 333–42.
HELMER, OLAF, and OPPENHEIM, PAUL. [Degree] "A Syntactical Definition of Probability and of Degree of Confirmation," *Journal of Symbolic Logic,* X (1945), 25–60.
HELMHOLTZ, HERMANN VON. [Zählen und Messen] "Zählen und Messen," in VON HELMHOLTZ, *Schriften zur Erkenntnistheorie.* Herausgegeben und erläutert von PAUL HERTZ and MORITZ SCHLICK. Berlin, 1921.

Bibliography

HEMPEL, CARL G. [Geometry] "Geometry and Empirical Science," *American Mathematical Monthly*, LII (1945), 7–17. Reprinted in FEIGL and SELLARS [Readings].
———. [Emp. Crit.] "Problems and Changes in the Empiricist Criterion of Meaning," *Revue internationale de philosophie*, No. 11 (1950), pp. 41–63.
———. [Cogn. Signif.] "The Concept of Cognitive Significance: A Reconsideration," *Proceedings of the American Academy of Arts and Sciences*, LXXX, No. 1 (1951), 61–77.
HEMPEL, CARL G., and OPPENHEIM, PAUL. [Typusbegriff] *Der Typusbegriff im Lichte der neuen Logik*. Leiden, 1936.
———. [Explan.] "Studies in the Logic of Explanation," *Philosophy of Science*, XV (1948), 135–75.
HERTZ, PAUL. [Comments] Comments on counting and measurement included in VON HELMHOLTZ [Zählen und Messen].
HILBERT, DAVID. [Grundlagen] *Grundlagen der Geometrie*. 4th ed. Leipzig, 1913.
HÖLDER, O. [Quantität] "Die Axiome der Quantität und die Lehre vom Mass," *Ber. d. Sächs. Gesellsch. d. Wiss., math.-phys. Klasse*, 1901, pp. 1–64.
HOGBEN, LANCELOT. [Interglossa] *Interglossa*. Penguin Books, 1943.
HULL, C. L. [Behavior] *Principles of Behavior*. New York, 1943.
———. [Int. Var.] "The Problem of Intervening Variables in Behavior Theory," *Psychological Review*, L (1943), 273–91.
HULL, C. L.; HOVLAND, C. I.; ROSS, R. T.; HALL, M.; PERKINS, D. T.; and FITCH, F. B. [Rote Learning] *Mathematico-deductive Theory of Rote Learning*. New Haven, 1940.
HUTCHINSON, G. EVELYN. [Biology] "Biology," *Encyclopaedia Britannica* (1948).
HUXLEY, JULIAN. [New Syst.] "Towards the New Systematics," in JULIAN HUXLEY (ed.), *The New Systematics*, pp. 1–46. Oxford, 1940.
KAPLAN, A. [Def. and Spec.] "Definition and Specification of Meaning," *Journal of Philosophy*, XLIII (1946), 281–88.
KOCH, SIGMUND. [Motivation] "The Logical Character of the Motivation Concept," *Psychological Review*, XLVIII (1941), 15–38 and 127–54.
LASSWELL, HAROLD D., and KAPLAN, ABRAHAM. [Power and Soc.] *Power and Society*. New Haven, 1950.
LAZARSFELD, PAUL F., and BARTON, ALLEN H. [Qual. Meas.] "Qualitative Measurement in the Social Sciences: Classification, Typologies, and Indices," in DANIEL LERNER and HAROLD D. LASSWELL (eds.), *The Policy Sciences*, pp. 155–92. Stanford, Calif., 1951.
LENZEN, VICTOR F. [EI5] *Procedures of Empirical Science*. EI5. Chicago, 1938.
LEWIS, C. I. [Analysis] *An Analysis of Knowledge and Valuation*. La Salle, Ill., 1946.
LUNDBERG, GEORGE A. [Foundations] *Foundations of Sociology*. New York, 1939.
———. [Measurement] "The Measurement of Socioeconomic Status," *American Sociological Review*, V (1940), 29–39.

Bibliography

———. [Definitions] "Operational Definitions in the Social Sciences," *American Journal of Sociology*, XLVII (1941–42), 727–43.

MACCORQUODALE, KENNETH, and MEEHL, PAUL E. [Distinction] "On a Distinction between Hypothetical Constructs and Intervening Variables," *Psychological Review*, LV (1948), 95–107.

MALINOWSKI, BRONISLAW. [Dynamics] *The Dynamics of Culture Change*. Ed. PHYLLIS M. KABERRY. New Haven, 1945.

MARGENAU, HENRY. [Reality] *The Nature of Physical Reality*. New York, 1950.

MAYR, ERNST. [Systematics] *Systematics and the Origin of Species*. New York, 1942.

MERTON, ROBERT K. [Social Theory] *Social Theory and Social Structure*. Glencoe, Ill., 1949.

MORRIS, CHARLES. [EI2] *Foundations of the Theory of Signs*. EI2. Chicago, 1938.

NAGEL, ERNEST. [Log. of Meas.] *On the Logic of Measurement*. (Thesis, Columbia University, 1931.) New York, 1930.

———. [Measurement] "Measurement," *Erkenntnis*, II (1931), 313–33.

———. [Geometry] "The Formation of Modern Conceptions of Formal Logic in the Development of Geometry," *Osiris*, VII (1939), 142–224.

———. [EI6] *Principles of the Theory of Probability*. EI6. Chicago, 1939.

———. [Op. An.] "Operational Analysis as an Instrument for the Critique of Linguistic Signs," *Journal of Philosophy*, XXXIX (1942), 177–89.

———. [Reduction] "The Meaning of Reduction in the Natural Sciences," in ROBERT C. STAUFFER (ed.), *Science and Civilization*. Madison, Wis., 1949.

NEUMANN, JOHN VON, and MORGENSTERN, OSKAR. [Games] *Theory of Games and Economic Behavior*. 2d ed. Princeton, 1947.

NORTHROP, F. S. C. [Logic] *The Logic of the Sciences and the Humanities*. New York, 1947.

———. [Einstein] "Einstein's Conception of Science," in P. A. SCHILPP (ed.), *Albert Einstein: Philosopher-Scientist*, pp. 387–408. Evanston, Ill., 1949.

OGBURN, WILLIAM F. [Social Change] *Social Change*. New York, 1922.

OGBURN, WILLIAM F., and NIMKOFF, MEYER F. [Sociology] *Sociology*. New York, 1940.

PAP, ARTHUR. [Anal. Philos.] *Elements of Analytic Philosophy*. New York, 1949.

PEANO, GIUSEPPE. [Définitions] "Les Définitions mathématiques," *Bibliothèque du Congrès International de Philosophie* (Paris), III (1901), 279–88.

POPPER, KARL. [Forschung] *Logik der Forschung*. Wien, 1935.

QUINE, W. V. [Convention] "Truth by Convention," in *Philosophical Essays for A. N. Whitehead*, pp. 90–124. New York, 1936. Reprinted in FEIGL and SELLARS [Readings].

———. [Math. Logic] *Mathematical Logic*. New York, 1940.

———. [Dogmas] "Two Dogmas of Empiricism," *Philosophical Review*, XL (1951), 20–43.

Bibliography

REESE, THOMAS W. [Measurement] *The Application of the Theory of Physical Measurement to the Measurement of Psychological Magnitudes, with Three Experimental Examples.* "Psychological Monographs," Vol. LV, No. 3 (1943).
REICHENBACH, HANS. [Axiomatik] *Axiomatik der relativistischen Raum-Zeit-Lehre.* Braunschweig, 1924.
———. [Raum-Zeit-Lehre] *Philosophie der Raum-Zeit-Lehre.* Berlin, 1928.
———. [Experience] *Experience and Prediction.* Chicago, 1938.
———. [Quantum Mechanics] *Philosophic Foundations of Quantum Mechanics.* Berkeley and Los Angeles, 1944.
———. [Logic] *Elements of Symbolic Logic.* New York, 1947.
———. [Probability] *Theory of Probability.* Berkeley and Los Angeles, 1949.
———. [Rise] *The Rise of Scientific Philosophy.* Berkeley and Los Angeles, 1951.
ROBINSON, RICHARD. [Definition] *Definition.* Oxford, 1950.
RUSSELL, BERTRAND. [Math. Philos.] *Introduction to Mathematical Philosophy.* 2d ed. London, 1920.
———. [Principles] *Principles of Mathematics.* 2d ed. New York, 1938.
SARGENT, S. STANSFELD, and SMITH, MARIAN W. (eds.). [Cult. and Pers.] *Culture and Personality: Proceedings of an Inter-disciplinary Conference Held under Auspices of the Viking Fund, November 7 and 8, 1947.* New York, 1949.
SCHLICK, MORITZ. [Erkenntnislehre] *Allgemeine Erkenntnislehre.* 2d ed. Berlin, 1925.
SHELDON, W. H. [Physique] *The Varieties of Human Physique.* With the collaboration of S. S. STEVENS and W. B. TUCKER. New York and London, 1940.
———. [Temperament] *The Varieties of Temperament.* With the collaboration of S. S. STEVENS. New York and London, 1945.
SOMMERHOFF, G. [Analyt. Biol.] *Analytical Biology.* London, 1950.
SPENCE, KENNETH W. [Theory Construction] "The Nature of Theory Construction in Contemporary Psychology," *Psychological Review*, LI (1944), 47–68.
———. [Learning] "Theoretical Interpretations of Learning," in S. S. STEVENS (ed.), *Handbook of Experimental Psychology*, pp. 690–729. New York and London, 1951.
STEVENS, S. S. [Loudness] "A Scale for the Measurement of a Psychological Magnitude: Loudness," *Psychological Review*, XLIII (1936), 405–16.
———. [Scales] "On the Theory of Scales of Measurement," *Science*, CIII (1946), 677–80.
———. [Math., Meas., and Ps.] "Mathematics, Measurement, and Psychophysics," in S. S. STEVENS (ed.), *Handbook of Experimental Psychology*, pp. 1–49. New York and London, 1951.
STEVENS, S. S., and DAVIS, H. [Hearing] *Hearing: Its Psychology and Physiology.* New York, 1938.
STEVENS, S. S., and VOLKMANN, J. [Pitch] "The Relation of Pitch to Frequency: A Revised Scale," *American Journal of Psychology*, LIII (1940), 329–53.

Bibliography

SUPPES, PATRICK. [Ext. Quant.] "A Set of Independent Axioms for Extensive Quantities," *Portugaliae Mathematica*, X, Fasc. 4 (1951), 163–72.

TARSKI, ALFRED. [Logic] *Introduction to Logic and to the Methodology of Deductive Sciences*. New York, 1941.

———. [Truth] "The Semantic Conception of Truth," *Philosophy and Phenomenological Research*, IV (1943–44), 341–75. Reprinted in FEIGL and SELLARS [Readings].

THURSTONE, L. L. [Reliability] *The Reliability and Validity of Tests*. Ann Arbor, Mich., 1932.

———. [Vectors] *The Vectors of Mind*. Chicago, 1935.

——— [Abilities] *Primary Mental Abilities*. "Psychometric Monographs," No. 1. Chicago, 1943.

———. [Analysis] *Multiple-Factor Analysis*. Chicago, 1947.

———. [Methods] "Psychophysical Methods," in T. G. ANDREWS (ed.), *Methods of Psychology*, chap. v. New York and London, 1948.

TOLMAN, E. C. [Op. Behav.] "Operational Behaviorism and Current Trends in Psychology," *Proceedings of the Twenty-fifth Anniversary Celebration of the Inauguration of Graduate Study, Los Angeles, University of Southern California, 1936*, pp. 89–103.

WALD, ABRAHAM. [Statist. Inf.] *On the Principles of Statistical Inference*. Notre Dame: University of Notre Dame, 1942.

WALKER, A. G. [Foundations] "Foundations of Relativity: Parts I and II," *Proceedings of the Royal Society, Edinburgh*, LXII (1943–49), 319–35.

WHITE, MORTON G. [Analytic] "The Analytic and the Synthetic: An Untenable Dualism," in S. HOOK (ed.), *John Dewey: Philosopher of Science and of Freedom*, pp. 316–30. New York, 1950.

WOLFLE, DAEL. [Factor Analysis] *Factor Analysis to 1940*. Chicago, 1940.

WOODGER, J. H. [Ax. Meth.] *The Axiomatic Method in Biology*. Cambridge, England, 1937.

———. [EII5] *The Technique of Theory Construction*. EII5. Chicago, 1939.

WOODROW, H. [Laws] "The Problem of General Quantitative Laws in Psychology," *Psychological Bulletin*, XXXIX (1942), 1–27.

Q
175
I 58
v.2 #7
c.2

MAR 16 1973